D1252787

70.1 S617
D PSYCHE
/SINNOTT, E
961 .00 FV
00 042401 30014
s Community College

570.1 S617 C.
SINNOTT
CELL AND PSYCHE
FV
.95
WITHDRAWN

CELL AND PSYCHE

(*continued on next page*)

W. H. Walsh — PHILOSOPHY OF HISTORY: *An Introduction* TB/1020
W. Lloyd Warner — SOCIAL CLASS IN AMERICA: *The Evaluation of Status* TB/1013
Alfred N. Whitehead — PROCESS AND REALITY: *An Essay in Cosmology* TB/1033

HARPER TORCHBOOKS / The Science Library

[Selected titles]

Angus d'A. Bellairs — REPTILES: *Life History, Evolution, and Structure.* Illus. TB/520
L. von Bertalanffy — PROBLEMS OF LIFE: *An Evaluation of Modern Biological and Scientific Thought* TB/521
David Bohm — CAUSALITY & CHANCE IN MODERN PHYSICS TB/536
R. B. Braithwaite — SCIENTIFIC EXPLANATION TB/515
P. W. Bridgman — THE NATURE OF THERMODYNAMICS TB/537
Louis de Broglie — PHYSICS AND MICROPHYSICS. Foreword by Albert Einstein TB/514
J. Bronowski — SCIENCE AND HUMAN VALUES TB/505
A. J. Cain — ANIMAL SPECIES AND THEIR EVOLUTION. Illus. TB/519
T. G. Cowling — MOLECULES IN MOTION TB/516
A. C. Crombie, *ed.* — TURNING POINTS IN PHYSICS TB/535
W. C. Dampier, *ed.* — READINGS IN THE LITERATURE OF SCIENCE. Illus. TB/512
H. Davenport — THE HIGHER ARITHMETIC TB/526
W. H. Dowdeswell — ANIMAL ECOLOGY TB/543
W. H. Dowdeswell — THE MECHANISM OF EVOLUTION. Illus. TB/527
C. V. Durell — READABLE RELATIVITY. Foreword by Freeman J. Dyson TB/530
Arthur Eddington — SPACE, TIME AND GRAVITATION TB/510
Alexander Findlay — CHEMISTRY IN THE SERVICE OF MAN. Illus. TB/524
Gottlob Frege — THE FOUNDATIONS OF ARITHMETIC. Trans. by J. L. Austin TB/534
R. W. Gerard — UNRESTING CELLS TB/541
C. Judson Herrick — THE EVOLUTION OF HUMAN NATURE TB/545
Max Jammer — CONCEPTS OF SPACE. Foreword by Albert Einstein TB/533
David Lack — DARWIN'S FINCHES TB/544
D. E. Littlewood — THE SKELETON KEY OF MATHEMATICS TB/525
J. E. Morton — MOLLUSCS: *An Introduction to their Form and Functions.* Illus. TB/529
J. R. Partington — A SHORT HISTORY OF CHEMISTRY. Illus. TB/522
H. T. Pledge — SCIENCE SINCE 1500: *A Short History of Mathematics, Physics, Chemistry, and Biology.* Illus. TB/506
John Read — A DIRECT ENTRY TO ORGANIC CHEMISTRY. Illus. TB/523
O. W. Richards — THE SOCIAL INSECTS TB/542
George Sarton — ANCIENT SCIENCE AND MODERN CIVILIZATION TB/501
Paul A. Schilpp, *ed.* — ALBERT EINSTEIN: *Philosopher-Scientist: Vol. I,* TB/502; *Vol. II,* TB/503
P. M. Sheppard — NATURAL SELECTION AND HEREDITY. Illus. TB/528
Edmund W. Sinnott — CELL AND PSYCHE TB/546
L. S. Stebbing — A MODERN INTRODUCTION TO LOGIC TB/538
O. G. Sutton — MATHEMATICS IN ACTION. Foreword by James R. Newman. Illus. TB/518
Stephen Toulmin — THE PHILOSOPHY OF SCIENCE: *An Introduction* TB/513
A. G. Van Melsen — FROM ATOMOS TO ATOM: *The History of the Concept* Atom TB/517
Friedrich Waismann — INTRODUCTION TO MATHEMATICAL THINKING TB/511
W. H. Watson — ON UNDERSTANDING PHYSICS TB/507
G. J. Whitrow — THE STRUCTURE AND EVOLUTION OF THE UNIVERSE TB/504
Edmund Whittaker — HISTORY OF THE THEORIES OF AETHER AND ELECTRICITY: *Vol. I, Classical Theories,* TB/531; *Vol. II, Modern Theories,* TB/532
A. Wolf — HISTORY OF SCIENCE, TECHNOLOGY AND PHILOSOPHY IN THE 16TH AND 17TH CENTURIES. Illus. *Vol. I,* TB/508; *Vol. II,* TB/509
A. Wolf — HISTORY OF SCIENCE, TECHNOLOGY, AND PHILOSOPHY IN THE 18TH CENTURY. Illus. *Vol. I,* TB/539, *Vol. II,* TB/540

HARPER TORCHBOOKS / The Cloister Library

[Selected titles]

Tor Andrae — MOHAMMED: *The Man and His Faith* TB/62
Augustine/Przywara — AN AUGUSTINE SYNTHESIS TB/35
Roland H. Bainton — THE TRAVAIL OF RELIGIOUS LIBERTY TB/30
C. K. Barrett, *ed.* — THE NEW TESTAMENT BACKGROUND: *Selected Documents* TB/86
Karl Barth — DOGMATICS IN OUTLINE TB/56
Karl Barth — THE WORD OF GOD AND THE WORD OF MAN TB/13
Nicolas Berdyaev — THE BEGINNING AND THE END TB/14
Nicolas Berdyaev — THE DESTINY OF MAN TB/61
J. H. Breasted — DEVELOPMENT OF RELIGION AND THOUGHT IN ANCIENT EGYPT TB/57
Martin Buber — ECLIPSE OF GOD TB/12
Martin Buber — MOSES: *The Revelation and the Covenant* TB/27
Martin Buber — THE PROPHETIC FAITH TB/73

(continued on next page)

Martin Buber	TWO TYPES OF FAITH TB/75
R. Bultmann, et al.	KERYGMA AND MYTH ; ed. by H. W. Bartsch TB/80
Jacob Burckhardt	CIVILIZATION OF THE RENAISSANCE IN ITALY. Intro. by B. Nelson and C. Trinkaus. Illus. Ed. *Vol. I,* TB/40 ; *Vol. II,* TB/41
Emile Cailliet	PASCAL : *The Emergence of Genius* TB/82
Edward Conze	BUDDHISM : *Its Essence and Development* TB/58
Frederick Copleston	MEDIEVAL PHILOSOPHY TB/76
F. M. Cornford	FROM RELIGION TO PHILOSOPHY TB/20
G. G. Coulton	MEDIEVAL FAITH AND SYMBOLISM TB/25
G. G. Coulton	FATE OF MEDIEVAL ART IN RENAISSANCE & REFORMATION TB/26
H. G. Creel	CONFUCIUS AND THE CHINESE WAY TB/63
Adolf Deissmann	PAUL : *A Study in Social and Religious History* TB/15
C. H. Dodd	THE AUTHORITY OF THE BIBLE TB/43
Johannes Eckhart	MEISTER ECKHART : A Modern Translation TB/8
Mircea Eliade	COSMOS AND HISTORY : *The Myth of the Eternal Return* TB/50
Mircea Eliade	THE SACRED AND THE PROFANE TB/81
Morton S. Enslin	CHRISTIAN BEGINNINGS TB/5
Morton S. Enslin	THE LITERATURE OF THE CHRISTIAN MOVEMENT TB/6
G. P. Fedotov	THE RUSSIAN RELIGIOUS MIND TB/70
Ludwig Feuerbach	THE ESSENCE OF CHRISTIANITY. Intro. by Karl Barth TB/11
Harry E. Fosdick	A GUIDE TO UNDERSTANDING THE BIBLE TB/2
Henri Frankfort	ANCIENT EGYPTIAN RELIGION : An Interpretation TB/77
Sigmund Freud	ON CREATIVITY AND THE UNCONSCIOUS TB/45
Maurice Friedman	MARTIN BUBER : *The Life of Dialogue* TB/64
O. B. Frothingham	TRANSCENDENTALISM IN NEW ENGLAND : *A History* TB/59
Edward Gibbon	THE TRIUMPH OF CHRISTENDOM IN THE ROMAN EMPIRE TB/46
C. C. Gillispie	GENESIS AND GEOLOGY : *The Decades before Darwin* TB/51
Maurice Goguel	JESUS AND THE ORIGINS OF CHRISTIANITY I : *Prolegomena to the Life of Jesus* TB/65
Maurice Goguel	JESUS AND THE ORIGINS OF CHRISTIANITY II : *The Life of Jesus* TB/66
Edgar J. Goodspeed	A LIFE OF JESUS TB/1
H. J. C. Grierson	CROSS-CURRENTS IN 17TH CENTURY ENGLISH LITERATURE TB/47
William Haller	THE RISE OF PURITANISM TB/22
Adolf Harnack	WHAT IS CHRISTIANITY ? Intro. by Rudolf Bultmann TB/17
R. K. Harrison	THE DEAD SEA SCROLLS : *An Introduction* TB/84
Edwin Hatch	THE INFLUENCE OF GREEK IDEAS ON CHRISTIANITY TB/18
Friedrich Hegel	ON CHRISTIANITY : *Early Theological Writings* TB/79
Karl Heim	CHRISTIAN FAITH AND NATURAL SCIENCE TB/16
F. H. Heinemann	EXISTENTIALISM AND THE MODERN PREDICAMENT TB/28
S. R. Hopper, *ed.*	SPIRITUAL PROBLEMS IN CONTEMPORARY LITERATURE TB/21
Johan Huizinga	ERASMUS AND THE AGE OF REFORMATION. Illus. TB/19
Aldous Huxley	THE DEVILS OF LOUDUN TB/60
Flavius Josephus	THE GREAT ROMAN-JEWISH WAR, with *The Life of Josephus* TB/74
Immanuel Kant	RELIGION WITHIN THE LIMITS OF REASON ALONE TB/67
Søren Kierkegaard	EDIFYING DISCOURSES : A Selection TB/32
Søren Kierkegaard	THE JOURNALS OF KIERKEGAARD : A Selection TB/52
Søren Kierkegaard	PURITY OF HEART TB/4
Alexandre Koyré	FROM THE CLOSED WORLD TO THE INFINITE UNIVERSE TB/31
Emile Mâle	THE GOTHIC IMAGE : *Religious Art in France of the 13th Century.* Illus. TB/44
T. J. Meek	HEBREW ORIGINS TB/69
H. Richard Niebuhr	CHRIST AND CULTURE TB/3
H. Richard Niebuhr	THE KINGDOM OF GOD IN AMERICA TB/49
Martin P. Nilsson	GREEK FOLK RELIGION TB/78
H. J. Rose	RELIGION IN GREECE AND ROME TB/55
Josiah Royce	THE RELIGIOUS ASPECT OF PHILOSOPHY TB/29
George Santayana	INTERPRETATIONS OF POETRY AND RELIGION TB/9
George Santayana	WINDS OF DOCTRINE *and* PLATONISM AND THE SPIRITUAL LIFE TB/24
F. Schleiermacher	ON RELIGION : *Speeches to Its Cultured Despisers* TB/36
H. O. Taylor	THE EMERGENCE OF CHRISTIAN CULTURE IN THE WEST : *The Classical Heritage of the Middle Ages* TB/48
P. Teilhard de Chardin	THE PHENOMENON OF MAN TB/83
D. W. Thomas, *ed.*	DOCUMENTS FROM OLD TESTAMENT TIMES TB/85
Paul Tillich	DYNAMICS OF FAITH TB/42
Ernst Troeltsch	THE SOCIAL TEACHING OF CHRISTIAN CHURCHES. Intro. by H. Richard Niebuhr. *Vol. I,* TB/71 ; *Vol. II,* TB/72
E. B. Tylor	THE ORIGINS OF CULTURE. Intro. by Paul Radin TB/33
E. B. Tylor	RELIGION IN PRIMITIVE CULTURE. Intro. by Paul Radin TB/34
Evelyn Underhill	WORSHIP TB/10
Johannes Weiss	EARLIEST CHRISTIANITY : *A History of the Period* A.D. *30–150.* Intro. by F. C. Grant. *Vol. I,* TB/53 ; *Vol. II,* TB/54
Wilhelm Windelband	A HISTORY OF PHILOSOPHY I : *Greek, Roman, Medieval* TB/38
Wilhelm Windelband	A HISTORY OF PHILOSOPHY II : *Renaissance, Enlightenment, Modern* TB/39

CELL AND PSYCHE

THE BIOLOGY OF PURPOSE

By

EDMUND W. SINNOTT

HARPER TORCHBOOKS / The Science Library
HARPER & ROW, PUBLISHERS
NEW YORK, EVANSTON, AND LONDON

CELL AND PSYCHE

Printed in the United States of America

This book was first published in 1950 by The University of North Carolina Press, and is reprinted by arrangement.

First HARPER TORCHBOOK edition published 1961

PREFACE TO THE TORCHBOOK EDITION

SINCE the publication of this little book in 1950, as the McNair Lectures at the University of North Carolina, the author has written two others, as well as a number of papers, on the same general theme. Though these elaborate the argument a little further, the essence of it is in *Cell and Psyche*. This is admittedly a speculation, but one based solidly on biological fact. It has been regarded as rather visionary and metaphysical by some people, but others have been attracted to it by the suggestion it offers for a better understanding of the ancient problem of how mind and body are related to each other. This problem is of such paramount importance, not only for a knowledge of what man really is but for the construction of a satisfying life philosophy, that any light thrown on it should be welcome.

The suggestion that man's physical life grows out of the basic goal-seeking and purposiveness found in all organic behavior and that this, in turn, is an aspect of the more general self-regulating and normative character evident in the development and activities of living organisms, is at least worth serious consideration. If we are to avoid a dualistic idea of man's nature and to construct a true monism that does not require the sacrifice of the significance of either mind or body, some such conception as this seems a reasonable means of doing so. It is to be hoped that the wider distribution now made possible for the present book may result in a more general consideration of this particular relationship between biology and philosophy.

E. W. S.

New Haven, Connecticut
April, 1961

CONTENTS

INTRODUCTION

In the clamor and confusion of our times one fact grows ever clearer—*beliefs* are important. One of the major problems with which men now are faced—perhaps, indeed, the most important one—is the wide disagreement which still exists in their fundamental philosophies. What course a man will follow, or a nation, is set in no small measure by his basic creed, by what he really thinks about the true nature of a human being—his personality, his freedom, his destiny, his relations to others and to the rest of the universe; by the judgments he makes as to what qualities and courses of action are admirable and should command his allegiance. These are not academic questions merely. They are ancient mysteries which long have troubled human hearts and seem today almost as far as ever from solution. The answer a man gives to them is the most significant thing that one can know about him. We may be tempted to underestimate the importance of these inner directives and turn instead to outer influences, to economic and social factors, as more decisive for our actions. But when we look at what the philosophy of Marx has done to set one-half the world against the other, at the basic divergence between the thinking of East and West, and at so many other differences in political and religious beliefs which now divide mankind, we can hardly doubt the profound practical import of men's philosophies. It is still true today that "as a man thinketh in his heart, so is he." In the minds of men are the most fateful battles fought. Against those ideologies we condemn, force in the end will fail. If our opponents cannot be convinced, or their ideas reconciled to ours, true peace will never come. And so today men everywhere are trying to formulate a satisfying body of convictions, a sound philosophy of life, in the hope that for a generation drifting on the ocean of uncertainty some anchor may take hold upon the bottom of eternal truth.

This is no new quest. Since the beginning of history men have pondered the deep questions of life and death, of beauty and truth, of good and evil. What brings so much confusion to their thinking now is the vast increase of scientific knowledge which has made nature much more difficult to understand and pulled down so many ancient pillars of belief. The answers confidently given a century ago and accepted then by almost everyone are seriously challenged today. The universe is a vastly bigger place than our grandfathers thought it was. The earth is far older than the 6,000 years allotted it by Bishop Ussher. Milton's account of the creation of living things in the seventh book of *Paradise Lost* is still fine poetry but quite inadequate as a scientific description of biological events. Man himself has a much longer and more complicated history than that with which a strict interpretation of the first chapter of Genesis would provide him. Even matter, the physicists tell us, which to the beginning of our own century retained its comforting solidity, should now be looked upon as largely empty space. The atom itself is no longer a little material pellet but has almost vanished into a series of electric charges, waves, and probabilities, no longer understandable except by mathematics. The universe is exploding. Space is curved. Light waves can be bent. Form and mass depend on speed of motion. The physical world of the nineteenth century with its reassuring certainties is gone forever, and we have not yet learned how to find our way about in the new one which has come to take its place.

It is not surprising, therefore, that men of science have become involved in an attempt to interpret their findings in philosophical terms and to bring order into our knowledge of the world. Eddington, Jeans, Schrödinger, Du Noüy, Sherrington, Needham, J. S. Haldane, Henderson, Julian Huxley, and many others are familiar names in this field today, as were those of T. H. Huxley, Haeckel, Driesch, and Jaques Loeb a half century ago. The conclusions of these earnest philosophic laymen are often open to criticism at the hands of those more skillful in the craft, and doubtless they have written much bad philosophy. As Professor Joad puts it, "When the scientist leaves his laboratory and speculates about the universe as a whole, the resultant conclusions are apt to tell us more about the scientist than about the universe."[1] But

[1] C. E. M. Joad, *Philosophical Aspects of Modern Science*, p. 339.

the wisest philosophers have failed to interpret the universe completely, and surely no aid in the accomplishment of so great a task should be despised. The fresh vision which the scientist can bring to these problems is stimulating, and the contribution which he makes to the philosophies of men may in the end prove more significant than all his triumphs in technology. It is in this belief that I shall here attempt to find some help toward a solution of a few of the great problems with which life confronts us by drawing upon the resources of biology, the science of life itself.

Whatever I shall say can hardly have much novelty, for biologists have long recognized the importance of their science for philosophy and have discussed these matters often and from many points of view. It is chiefly during the past hundred years, however, when biology has really come of age, that they have left the laboratory and the field from time to time to engage in the thrust and parry of philosophic argument.

Darwin's theory of Natural Selection and the subsequent wide acceptance of the fact of organic evolution wrought a complete revolution in our understanding of the origin of living things and of man himself and thus in our whole attitude toward nature. Not only did a literal interpretation of the traditional story of creation become impossible for anyone who understood biology and geology, but the new ideas challenged tradition on even more fundamental grounds than that of Biblical infallibility. Though Darwin himself showed little interest in the implications of his theory, some of the early protagonists of his ideas drew from them strong support for a philosophy of materialism and thus struck at the very basis of religion. Ernst Haeckel, Darwin's first great supporter on the continent, battled with enthusiastic vigor against the ancient ideas of immortality, free will, and the existence of a God. Life for him was simply a complex chemical phenomenon associated with the compounds of carbon and had originated by evolution from inorganic nature. Mind was merely the result of chemical changes in living stuff, the soul a fiction, and man himself no more than part of a rigidly determined, planless universe. All nature was one, and matter its sole foundation. This extreme and uncompromising materialism found less favor in England and America. Darwin's biological theories led T. H. Huxley into wide discussions of philosophy and religion in which he adopted an essentially agnostic attitude, supporting the concept of universal causation and the supremacy of the scientific approach to truth. He broke sharply with orthodox Christianity, but refused to call himself a materialist.

But something more than evolutionary speculation was needed. The question of what sort of system a living thing is could readily be approached by experiment, and in the latter years of the nineteenth century the science of experimental embryology opened up a new and exciting chapter in biology and biological philosophy. The fertilized eggs and early developmental stages of many animals, notably some of the echinoderms and amphibia, can be studied under controlled conditions and manipulated in various ways. The results of work on such material were in many cases most surprising. When, for example, one of the first two cells of a tiny salamander embryo is destroyed, the remaining one grows into a whole individual, not a half, as one might expect. Two fertilized eggs induced to fuse by artificial means were found to produce one animal instead of two. A mass of evidence of this sort made it very difficult to interpret development as due to the progressive parcelling out of "determiners" from an egg to the cells which originate from it, as Weismann and others had suggested. It soon became evident that every cell of many embryos, at least in the early stages, is able, if isolated, to produce a whole animal. The implications of this remarkable fact were not overlooked, and Driesch was quick to point out the difficulties involved in imagining a machine capable of being cut up into an indefinite number of pieces each of which could restore the whole machine again. For this and other reasons he concluded that a mechanistic explanation of development was impossible, and postulated the operation here of an *entelechy*, an extra-physical agent which in some unknown manner guides the course of development. Such vitalism has found little acceptance among biologists, although there are a few thinkers today who call themselves neo-vitalists. Man's mind, which has knowledge of all nature as its goal, will not readily accept defeat by admitting that there is something here which must lie beyond its power to understand. The grave problems which such studies in experimental embryology raise, however, are still far from solution.

Meanwhile the geneticists had not been idle. Mendel's laws were rediscovered, and by a brilliant series of inductions the physical basis of inheritance was shown to consist of a series of genes arranged in a constant pattern in the chromosomes of the nucleus. Since all cells of the body have the same number of chromosomes, they are very probably identical in hereditary constitution, and the question of how these similar cells cooperate to produce a complete individual with its specific form and its patterned differences is hard to understand. Developmental genetics, which emphasizes the mechanics of gene action, thus faces essentially the same problem as does experimental embryology—that of how the orderly control of development is accomplished. To understand the means by which thousands of genes—presumably protein molecules—in every cell can so guide the chemical activities of protoplasm that an *organism* is produced, is very difficult. J. S. Haldane went so far as to say that "the mechanistic theory of heredity is not merely unproven, it is impossible. It involves such absurdities that no intelligent person who has thoroughly realized its meaning and implications can continue to hold it."[2]

But many were not so pessimistic as this about the ability of science to explain an organism in terms of mechanism. By the turn of the century physiologists had

[2] J. S. Haldane, *Mechanism, Life and Personality*, p. 58.

made much progress in the physico-chemical analysis of the activities of living things. Many vital processes could by then be imitated outside the body, and Jaques Loeb, a vigorous champion of the mechanistic interpretation of life, was already hopefully predicting the artificial synthesis of living protoplasm in the near future. Biochemistry, continuing to apply to the analysis of living things the chemical knowledge already gained in the laboratory, has now become one of the most active and promising fields of scientific inquiry, and physiologists delve hopefully into the maze of proteins, nucleic acids, enzyme systems, and hormones of which protoplasm is composed to learn much about its complex structure and activities. Facts which they find can be interpreted in chemical terms, and there is no suggestion here of any mystical or superphysical agent. The possibilities before this growing discipline are great, and its practitioners, undaunted by the unsolved problems which they must still encounter, believe that biochemistry holds the key which will unlock all the secrets of life. Such a belief gives support to a conception of the organism as a physico-chemical mechanism and thus to a materialistic approach to life's problems. This is the attitude of many biologists today and is expressed in its more extreme form by writers like Lancelot Hogben.

A number of biological philosophers, impressed by the enormous complexity of life and the difficulties of explaining it in mechanistic terms, have endeavored in various ways to avoid the extremes of both materialism and vitalism by finding some middle position which can accept the results of physiology and genetics but still find room for freedom, purpose, and value. Notable among them are those who believe that the living *organism* introduces a new concept quite different from that of a physical mechanism or a mystical entelechy. Unless we study living things as integrated systems with characteristics and laws of their own, say these organicists, we shall never understand what life really is. This important idea, which would set up biology as an independent science and not merely a complicated kind of physics or chemistry, has found favor with many. Notable among them are J. S. Haldane, Bertalanffy, Ritter, and Smuts, who differ in the details of their theories but agree that the organism is the heart of the problem. As to just how this system is set up and persists, however, they have few suggestions to offer.

A somewhat different philosophical approach is made by those who advocate what is called *emergent evolution*. They accept the truth of experimental determinism—that similar conditions will always be followed by similar consequences—since without this science would lose meaning, but they point out that as evolution progresses, and as *new* conditions arise which have never been present before, these will invariably be followed by new consequences. Such consequences, they believe, cannot be predicted from a knowledge of the conditions alone. This concept, first elaborated in detail by Lloyd Morgan, does away with the necessity of believing in a rigidly mechanical universe. It allows opportunity for novelties, for new phenomena and new principles; even for freedom, since antecedent conditions are never precisely the same. Life can be regarded as such an "emergent." Although it follows laws, these are laws of its own and not necessarily predictable from or determined by those of physics and chemistry. This theory has found favor with a number of biologists and has the obvious advantage of providing autonomy for life without interfering with physical determinism, as vitalism does, but it also possesses certain logical difficulties, and its final contributions to biological philosophy are still uncertain.

But while the physiologists have been busily at work applying the techniques and concepts of chemistry to an analysis of life processes, the physicists have been revolutionizing the whole conception of matter itself, and with most surprising results. Relativity and quantum mechanics have greatly altered our ideas of the physical universe. In the world of Planck and Heisenberg and Einstein the confident philosophical conclusions of the mechanists, however certain their experimental results may be, begin to sound a little naïve. The matter on which materialism is based has become so tenuous that it seems hardly able to support the tough-minded philosophy of a Haeckel or a Loeb, and the men who approach philosophy from the physical sciences are much less dogmatic today than the biologists, as those who read Eddington and Jeans and Schrödinger well know.

A number of distinguished physicists have speculated

about the nature of life. Nils Bohr, for example, one of the notable pioneers in atomic theory, proposes another explanation for the gap between life and the non-living, which to many still seems so unbridgeable. He is impressed with the difficulty of reconciling classical mechanics with the newer quantum theories and suggests that they may be parallel and complementary ways of looking at the universe. Each has its own laws, each provides an orderly system of scientific facts, but neither can be derived from the other. He sees a resemblance here to the relation between physics and biology and postulates a similar *complementarity* between living systems and lifeless ones. Each, he believes, is autonomous and has its own specific rules and principles, but the attempt to derive life from matter, to regard it simply as a complex physico-chemical system, he is inclined to think is a difficult or even impossible quest. "The existence of life," he says, "must be considered as an elementary fact that cannot be explained, but must be taken as a starting point in biology, in a similar way as the quantum of action, which appears as an irrational element from the point of view of classical mechanical physics, taken together with the existence of elementary particles, forms the foundation of atomic physics."[3] This idea has received a respectful hearing, but it violates our deep desire to bring all nature into a single system, to make the universe truly *one*. Perhaps such a program is too ambitious, but scientists will not abandon it until they are forced to do so.

[3] N. Bohr, "Light and Life," *Nature*, 131 (1933), 421-23, 457-59.

Another notable physicist, Erwin Schrödinger, looks for the solution of this problem in still undiscovered laws of physics. "We must not be discouraged," he says, "by the difficulty of interpreting life by the ordinary laws of physics. For that is just what is to be expected from the knowledge we have gained of the structure of living matter. We must be prepared to find a new type of physical law prevailing in it."[4] Science is young, and surely we have not yet discovered all the laws which govern the universe. A century from now our present difficulties may well have been removed, though doubtless others, equally serious, will by then have arisen.

So swing the tides of theory to and fro. From the vast amount of study and thought which have been given to the problem only one conclusion can be drawn with certainty today—we still are a long way from understanding what life really is. Man has not yet gone far enough along the roads of scientific discovery and of philosophical insight to be able to answer the thronging questions which life raises. All that each of us can do is to adopt as a working hypothesis that one which seems most likely to give opportunity for further progress. We may become vitalists and undertake the almost hopeless task of learning something about entelechies. Or, following the now popular course, we may work as strict mechanists and endeavor to interpret all life, from bottom to top, in terms solely of matter and of energy. If neither extreme satisfies, we may follow some middle way like emergent

[4] E. Schrödinger, *What Is Life?*, p. 80.

evolution or complementarity, studying life by and for itself without trying to tie it to lifeless nature. Or finally, admitting that on the basis of our present knowledge of both facts and laws the problems of life seem insoluble, we may push out on every front in search of fresh facts, of adventurous hypotheses, of new principles as science advances through the years, which may give us a clue as to the nature of living organisms.

In this slender volume I shall add nothing to our store of scientific knowledge, nor can I hope to say anything very new about biology or its theories of life. Whatever novelty the present discussion achieves and whatever merit it may possess will come from its attempt to bring some of the concepts of biology in the narrower sense closer to those of psychology and of philosophy to help end what McDougall calls "the intolerably absurd state of affairs hitherto obtaining; namely, two sciences of the functioning of organisms, on the one hand mechanical biology, on the other psychology; two sciences completely out of touch with one another; the one ignoring the mental life of men and animals, the other trying vainly to relate it intelligibly to the bodily life."[5] From such a synthesis there may emerge a few ideas useful in answering the difficult questions with which we began our discussion. It is not, of course, the problems of theoretical biology as such which chiefly interest man or are significant for his welfare, but the higher aspects of man's life which emerge from these. Such are problems of biology

[5] William McDougall, *The Riddle of Life*, p. 265.

in its broad sense but of biology at a much higher level than that of the laboratory. If we can bring these loftier matters down to their common protoplasmic denominators and find some problem which is basic to all of them, we shall help to clarify the great objectives which we seek and to bring unity into the search. If it is possible to disentangle from the throng of biological and psychological ideas, often seemingly so irrelevant and confusing, one main issue—the key log in the jam—we shall then be able to focus attention and investigation on this point, undistracted by minor and irrelevant details, and thus more hopefully approach our goal. I believe that such a basic problem does exist, that it touches the life of the mind and spirit as well as of the body, and that upon its solution depends not only our understanding of life in strictly biological terms but our ability to answer some of the deep questions which men so long have asked. To discover it and to explore its philosophical implications is the purpose of the following pages.

CHAPTER I

ORGANIZATION, THE DISTINCTIVE CHARACTER OF ALL LIFE

THE UNIVERSE is turning out to be a far more surprising place than our grandfathers ever dreamed. The more we learn of it, the wider grows the realm of the unknown. Science, like Hercules, is coping with a Hydra and finds that for every problem which is solved, two new ones rise at once to take its place. "An addition to knowledge," says Eddington, "is won at the expense of an addition to ignorance. It is hard to empty the well of Truth with a leaky bucket."[1]

In the last generation physics and chemistry and astronomy have completely rebuilt our old ideas about the world of nature. Ancient solidities and certainties have disappeared. Matter and energy and space and time have taken on quite other aspects and seem to be subject to analysis at last only by mathematical subtleties. During this same half century the sciences of life have also made great progress, especially in the application of physical and chemical knowledge and techniques to biological problems; but there has been no such revolution here as that which shook the physical sciences so profoundly.

[1] A. S. Eddington, *The Nature of the Physical World*, p. 229.

Biology is far more complex than they, and for it there has not yet come a formulation of the new and radical concepts which are necessary before life can truly be understood. Protoplasm still confronts us as the most formidable of enigmas.

But for all men life must nevertheless remain the ultimate problem. Around it, since we ourselves are living things, center those great questions which have always stirred mankind most deeply: on the lower level food, sex, race, and other problems of our animal nature; on the higher, those ultimate questions as to the place and significance of man in the universe, as to his personality and its destiny, his freedom, and the meaning for him of love and beauty, of virtue and aspiration, of what he calls his spirit and its communion with the universe outside of him. The structure of the atom, the size of space, and the theory of relativity interest a few, but rarely stir men deeply. No one goes to the barricades in defense of $E = mc^2$. But those more vital matters, which reach into our hearts as much as into our minds, have set wars ablaze and banners flying and poets singing and mystics praying since the dawn of history. These are all problems of *life*, and life is the ultimate mystery.

Any satisfying philosophy must deal with these questions, and to do so it must be rooted in the science of life itself, of life not only as we see it in man but as it is expressed in those far simpler organisms up and down the evolutionary scale. It is therefore biology in its widest sense, as the interpreter of life at every level, which will bring the richest offerings to philosophy. Tennyson's flower in the crannied wall, if we could really understand it, "root and all, and all in all," would indeed solve for us the final mysteries of God and man, for these are the mysteries of life itself.

What, then, can the biologist tell us about the curious phenomenon with which he deals? The nineteenth century produced the magnificent conception of life as dynamic, changing, ever moving forward; of the history of the world as the great stage on which the drama of organic evolution is being enacted. But it also established the equally important conception that life has its physical basis in that remarkable material system which is called *protoplasm*. Here in this aggregation of proteins—watery, formless, and flowing, deceptive in its visible simplicity but amazingly complex in its ultimate organization—are centered all the problems of living things. It is not greatly different chemically and physically in bacterium and orchid, in amoeba, arthropod, and man. Life is protoplasmic activity, and this is essentially the same from protozoan to primate. Man is not only cousin to all living things by blood-relationship, but is built of the very same stuff as they. It is not of dust or clay that we all are made, but of proteins and of nucleic acids.

The task of the biologist is therefore to understand this remarkable living material. From it are built the beautiful and intricate bodies of plants and animals; in it centers the control which regulates the activities of these exquisite mechanisms; and out of it come the alterations which make possible all evolutionary change. Early biologists believed that there must be some sort of soul or *anima* in every living thing, which governs it. A few, even in recent times, have been so much impressed with the complexity of protoplasmic activity, especially in its control of growth and development, that they adopt an essentially similar explanation and assume the existence of an entelechy or some other extra-physical agent which directs the activity of living stuff. Such a philosophy of vitalism, however, is now rarely asserted. Students of plant and animal physiology more commonly seek to explain in physical and chemical terms alone everything that goes on in protoplasm. The recent rapid growth of biochemistry has made it possible to analyze into relatively simple processes so many vital activities that this mechanistic view of biology has been greatly stimulated. It looks at life as simply a particularly complex series of physical and chemical reactions, no different fundamentally from those in any material system.

Protoplasm is a far more complicated affair, however, than biologists of a generation ago imagined it to be. An easy imitation, outside the organism, of some of the changes evident in living cells led them to the optimistic prediction that in a few years it would be possible even to synthesize protoplasm and produce a living thing. Such a triumph today seems farther away than ever. Physiologists have underestimated their protoplasmic opponent and have been obliged to withdraw, at least temporarily, from many advanced theoretical positions. Everything that we have learned about protoplasm in recent years testifies to a complexity in physical structure, chemical composition, and physiological activity within it far beyond that which its visible simplicity would lead us to expect; and when we realize that out of this remarkable stuff has come not only the protean plant and animal life of our globe but man himself, with the magnificent accomplishments and the sublime possibilities which are his, our respect for it should be profound. Protoplasm is a bridge anchored at one end in the simple stuff of chemistry and physics, but at the other reaching far across into the mysterious dominions of the human spirit.

A recognition of the magnitude of the problem which confronts them has far from discouraged biologists. For its solution they have enlisted the aid of their own clans—physiologists, morphologists, embryologists, geneticists, cytologists, microbiologists, and the rest—and have called in powerful allies from chemistry, physics, and mathematics. Their successes have been notable. The electron microscope has delved so deeply into protoplasmic structure that the genes themselves at last are visible. Some of the processes of metabolism, notably that of respiration, once thought to be fairly simple chemical exchanges, have been shown to involve many and complex steps and interactions and the mediation of a long series of enzymes. Growth and development in animals and plants are known to be affected by many chemical and physical factors—hormones, growth substances, organizers, bioelectric fields, light, temperature, and many others. Every living thing, even the humblest, is evidently a mechanism of the most remarkable and exquisite complexity.

What, we may ask, is the essential character of this mechanism, the quality that best distinguishes it? An obvious answer would be that it contains some substance or substances which make it what it is. This answer has often been given; and the increase in our knowledge of the chemical activities of living stuff and of the physiological importance of specific substances like the hormones has persuaded many biologists that the secret of life is indeed to be found in a persistent analysis of its biochemical behavior.

Others, however, who see the difficulty of this concept if it is carried very far have come to realize that it is not the *character* of the constituents of a living thing but the *relations* between them which are most significant. An organism is an *organized* system, each part or quality so related to all the rest that in its growth the individual marches on through a series of specific steps to a specific end or culmination, maintaining throughout its course a delicately balanced state of form and function which tends to restore itself if it is altered. This is the most important thing about it. E. B. Wilson in a famous passage said that "we cannot hope to comprehend the activities of the living cell by analysis merely of its chemical composition. . . . Modern investigation has, however, brought ever-increasing recognition of the fact that the cell is an *organic system*, and one in which we must recognize some kind of ordered structure or organization."[2] Woodger remarks that "biologists in their haste to become physicists, have been neglecting their business and trying to treat the organism not as an organism but as an aggregate. . . . If the concept of organization is of such importance as it appears to be it is something of a scandal that we have no adequate conception of it. The first duty of the biologist would seem to be to try and make clear this important concept. Some biochemists and physiologists . . . express themselves as though they really believed that if they concocted a mixture with the same chemical composition as what they call 'protoplasm' it would proceed to 'come to life.' This is the kind of nonsense which results from forgetting or being ignorant of organization."[3] Herbert Muller puts it well thus: "For the fundamental fact in biology, the necessary point of departure, is the organism. The cell is a chemical compound but more significantly a type of biological organization; the whole organism is not a mere aggregate but an architecture; the vital functions of growth, adaptation, reproduction—the final function of death—are not merely cellular but organic phenomena. Although parts and processes may be isolated for analytical purposes, they cannot be understood without reference to the dynamic, unified whole that is more than their sum. To say, for example, that a man is made up

[2] E. B. Wilson, *The Cell in Development and Inheritance*, p. 760.
[3] J. H. Woodger, *Biological Principles*, pp. 281, 290.

of certain chemical elements is a satisfactory description only for those who intend to use him as a fertilizer."[4]

Through all the complexity which it is the task of the biologist to analyze thus runs one fundamental fact common to every living thing: protoplasm *builds organisms.* It does not grow into indeterminate, formless masses of living stuff. The growth and activity shown by plants and animals are not random processes but are so controlled that they form integrated, coordinated, organized systems. The word *organism* is one of the happiest in biology, for it emphasizes what is now generally regarded as the most characteristic trait of a living thing, its *organization.* Here is the ultimate battleground of biology, the citadel which must be stormed if the secrets of life are to be understood. All else are outworks, easily open to energetic attack. But this central stronghold, we must ruefully admit, has thus far almost entirely resisted our best efforts to break down its walls.

Organization is evident in diverse processes, at many levels, and in varying degrees of activity. It is especially conspicuous in the orderly growth which every organism undergoes and which produces the specific forms so characteristic of life. A coniferous tree, for example, such as a spruce or pine, though a loosely integrated plant individual in comparison to most animals, has a definite form, and its parts show a close coordination with each other. In each year's growth the central shoot is vertical and continues the axis of the trunk. The several side

[4] Herbert J. Muller, *Science and Criticism*, p. 107.

shoots, not as long, spread out almost horizontally. A definite pattern for the crown of the tree thus develops, the apical shoot growing faster than the branches, but the ratio between the two remaining essentially constant so that a regular conical shape is produced. If the young "leader" or terminal shoot is removed, one of the laterals swings up to take its place. This and other evidence indicates that the orientation and the relative growth of these side shoots are in some way under the control of the terminal bud. Other buds govern the growth of particular parts or branches. The angles which these make with the trunk, the ratio of height to diameter in the trunk itself, the proportion of above-ground parts to the root system, and other measurable relationships tend to be maintained. Thus the whole tree is an organized system in which the character and amount of growth in one part is related to that in all the others so that a precise form is reached. Some of the agents involved in this control, notably the plant hormones, are known; but how they are distributed so precisely in space and time that such a coordinated system is produced we do not understand. The tree itself is the expression of this organizing control.

A still more tightly organized system is evident in the developing animal embryo. The fertilized egg of a salamander is cleft into two cells by a vertical wall, then into four as one would quarter an apple, then horizontally into eight, and so on and on. If to our vision these changes are speeded up by time-lapse photography we can witness how the tiny group of cells, through continued cleavage, forms a partly hollow, spherical body; how the upper portion grows down over the yolk mass; how at one point the sphere is pushed in to make the primitive mouth; how above this the puckered neural folds mark out the position of the spinal axis; how they grow over to meet and form the neural tube; how at the sides the primitive gills appear; and how, step after step, the embryo moves swiftly on to form the young larva from which the mature salamander grows. Here is no random process but a steady march, each event in step with the rest as though to a definite and predetermined end. One gets an impression of some unseen craftsman who knows what he is about and who molds the mass of growing cells according to a precise plan. The young salamander seems to go through, before our eyes, an active progress toward a destination in a way which suggests its later movements of behavior, and not a merely passive unfolding. Here seems to be the expression in development of a constantly operating control which from the start and through all its precise steps from egg to adult maintains the embryo as an organized system.

This strict coordinated progression in organic growth is everywhere manifest, though often in less dramatic ways. The very fact that living things, in their bodies and in the organs which constitute them, everywhere show constant and specific *forms*, is proof of this control. Form is simply the external and visible expression of the organizing activity of protoplasm and is thus perhaps the most distinctive characteristic of living things. As Needham has well said, the central problem of biology is the form problem. In a gourd fruit, for example, growth in length and in width proceed at different rates so that form, as indicated by the ratio of one dimension to another, is continually changing. What remains constant is the ratio of the growth rates. During the development of the fruit any two rates keep evenly in step with each other so that it is possible to predict the actual dimensions and the changing dimensional ratios, and thus the organic pattern, at any stage of growth.

In the light of these facts it is impressive to look under the microscope at a thin slice of an early stage in such a developing fruit. Here one sees hundreds of tiny cells which by their constant division cause the organ to grow. The planes of these divisions—the angle of each new partition wall which cuts a cell into two—are in all directions. Chaos here seems to reign. This is no chaos, however, but a cosmos, with events marching to a precise destination, for the growth in the various dimensions of the organ which results from these divisions is beautifully coordinated. Some integrating control must here be operating. It is the nature of this control, of this fundamental organizing activity, which still eludes us and which constitutes the most formidable problem of biology.

One could multiply indefinitely examples of this sort, since all development normally shows such organized behavior; but among the lowliest of fungi there is an instance of this so remarkable that it illuminates the whole problem. In one group of slime molds (the *Acrasiaceae*) the individuals are single cells, each a very tiny and quite independent bit of protoplasm resembling a minute amoeba. These feed on certain types of bacteria found in decaying vegetable matter and can readily be grown in the laboratory. They multiply by simple fission and in great numbers. When this has gone on for some time a curious change comes over the members of this individualistic society. They cease to feed, divide, and grow, but now begin to mobilize from all directions toward a number of centers, streaming in to each, as one observer describes it, like people running to a fire. Each center exerts its attractive influence over a certain limited region, and to it come some thousands of cells which form a small elongated mass a millimeter or two in length. These simple cells do not fuse, but each keeps its individuality and freedom of movement. The whole mass now begins to creep over the surface with a kind of undulating motion, almost like a chubby worm, until it comes to a situation relatively dry and exposed and thus favorable for spore formation, where it settles down and pulls itself together into a roundish body. Now begins a most curious bit of activity. Certain cells fasten themselves securely to the surface and there form collectively a firm disc. Others in the central axis of the mass become thick-walled and form the base of a vertical stalk. Still others, clambering upward over their comrades, dedicate themselves to the continued growth of the stalk.

Up this stalk swarms the main mass of cells until they have risen several millimeters from the surface. These cells, a majority of the ones which formed the original aggregate, now mobilize themselves into a spherical mass terminating the tenuous stalk, which itself remains anchored to the surface by the basal disc. In this terminal mass every cell becomes converted into a rounded, thick-walled spore which, drying out and blown away by the wind, may start a new colony of separate amoeba-like cells. In other species the structure is even more complex, for the ascending mass of cells leaves behind it groups of individuals which in turn form rosettes of branches, each branch terminating in a spore mass. In this process of aggregation, a group of originally identical individuals is organized into a system wherein each has its particular function and undergoes a particular modification, some cells to form the disc, others the stalk, and others serving as reproductive bodies.

Such an aggregation of distinct cellular individuals into an organized system may also be observed in certain sponges. The living part of the body of such animals, consisting of at least four different kinds of cells, can be broken up artificially and even passed through muslin, but the thoroughly disorganized mass of cells, if they are not injured in the process, will regroup themselves in proper positions and produce a whole animal again. In some respects even more remarkable is a process which takes place among many insects, where the tissues of the caterpillar are broken down during the pupal or cocoon stage into what appears to be a disorganized mass of "mush." Out of this unpromising material the entirely different tissues and organs of the adult insect are gradually mobilized, a metamorphosis indeed, and one of the enigmas of biology.

Such organizing behavior is somewhat different from that in most plants and animals since here all growth (increase in material) is finished before differentiation and development begin, but we can hardly doubt that the process which integrates this group of individuals or a mass of homogeneous material and transforms it into an organized biological society is the same as that which operates in the more familiar cases of growth and development by cell multiplication. In both there is the same orderly progression, the same close coordination of one part with the rest, the same march to a final goal. In both, to use Driesch's famous dictum, "The fate of a cell is a function of its position." In both, there is the same evidence of unifying control. Surely if we could understand what makes the tiny cells of a slime mold run together and build such a beautifully fashioned cell-state, where each is modified in a particular way which serves the whole, we should know much about the ultimate secret of life.

The evidence of biological organization from these examples of normal growth and development is greatly extended through studies by which these processes are experimentally modified, especially by removing certain parts of the growing body. When this happens the organism shows a remarkable ability to regenerate its lost parts and restore a normal whole. Thus a "cutting," removed from a plant, under proper conditions will produce a new root system and finally an entire individual with the normal proportion of root to shoot. Internal plant structures may also be restored. If a conducting bundle in the growing stem or leaf is severed, the two ends may be connected by the development, behind the cut, of a new bundle through the conversion of ordinary storage cells into specialized vascular ones.

Some of the most remarkable examples of regeneration occur in animal embryology. Where the egg of a sea urchin or a frog, for example, at the beginning of development divides into two cells, these may be separated from each other, and each, instead of producing *half* an individual, now grows into a *whole* one. The fate of each cell is now quite different from what it would have been if it had remained part of a two-celled embryo. By the reorganization of its material each regenerates a single whole animal. Such behavior, of which countless similar examples might be cited, is so difficult to explain on chemical or physical grounds that Driesch, less tough-minded than most biologists, was driven to assume the operation here of an entelechy or director.

Regeneration is common everywhere in young, growing organisms. The leg of a tadpole, snipped off, may be restored, or the eye of a crustacean. Mature animals also may regenerate, as in the familiar case of the angleworm in which, when the body is cut in two, the head end will form a new tail. Regenerative ability is by no means universal, however, and is lost in most adult individuals or structures. In less highly organized systems, like most plants, it persists in certain more embryonic parts. Many cases are known where a single cell, from the surface layer of a leaf or elsewhere, may be induced to start independent development and to form an entire new individual. The general conclusion, with all its far-reaching implications, seems justified that every cell, fundamentally and under proper conditions, is *totipotent*, or capable of developing by regeneration into a whole organism.

In all these cases of regeneration the molding, coordinating, organizing activity of living stuff is emphasized. Here, as in normal cases, the ultimate result, the goal toward which development seems to move, tends to be a single complete organism, whatever may have been the vicissitudes of its developmental history. The organizing ability of protoplasm thus shown so dramatically in the processes of growth and development has long excited the interest of biologists. To answer the problems which it poses is the task of the science of morphogenesis, which endeavors to mobilize evidence and techniques for their solution from most of the other biological disciplines and from the physical sciences, as well.

This same organizing control is evident not only in development but in the protoplasmic activities by which the life of the individual is maintained. Around a living creature is its unorganized material environment, a random mixture of many things. Certain of these, its food, are continually being pulled into the organism, where at once they lose their random character and are built into the organized structure of a living system. Every plant and animal thus acts as an incorporating center which brings organic order out of environmental disorder.

Such a living organism, however, is extraordinarily unstable and sensitive to external conditions. It is an open system, and matter is continuously passing into it and out of it. It is the seat of innumerable chemical and physical changes incident to vital activity. And yet the very continuance of its life depends on the maintenance of relatively constant conditions within it—of water content, acidity, oxygen supply, a definite concentration of certain specific substances, and many more. This is not merely an equilibrium, a balance between forces. It is what the physiologists call a "steady state," and to maintain it the expenditure of energy is required. Life *is* the maintenance of such a constant set of conditions, and death is the inevitable result of their dislocation. In such a complex and open system, the first requisite is evidently a means whereby the many activities are so regulated that the necessary balance is constantly restored as external and internal changes upset it, and the inevitable tendency toward disorganization is continually resisted. Here again, as in the processes of development, each part of the system must be closely tied to all the rest so that changes in one activity or in one region may be compensated by those in another. It is therefore very hard for a physiologist to study any one activity by itself, a fact which makes the practice of this science peculiarly difficult and has led to many erroneous conclusions. The particular level of physiological balance may change as development progresses, or as the environment is altered, but for each state or condition there is set up in the organism a norm or standard toward the maintenance of which its activity is constantly regulated.

The most conspicuous and best known of these physiological regulations are those in the higher animals, particularly the mammals, which must maintain a very constant internal environment. The precisely controlled bodily temperature of man and the warm-blooded animals is a common example of this. Equally important, though less familiar, is the maintenance of uniform concentrations of sugar and oxygen in the blood, and similar constancies. The control of these physiological processes is well described by Cannon in a notable book.[5] He proposes for this state of balance the term *homeostasis*. The way in which this is maintained under changing conditions, and the ability of the body to regulate its vital processes so very delicately, is surely one of the most remarkable phenomena displayed by living things.

Such regulations are familiar in man and the higher animals, where the mechanisms involved are chiefly the nervous system and glands of internal secretion. Living cells which are far less specialized, however, are also capable of such self-regulation. Among plants, for example, the hydrogen-ion concentration (acidity) in the

[5] W. B. Cannon, *The Wisdom of the Body*.

sap of cells of a given tissue is often very closely maintained despite external change. The concentration of various dissolved chemical substances in particular cells may also be kept very near to a given level under widely varying external concentrations. These are essentially the same sorts of regulations as in homeostasis but involve no nervous mechanism.

A remarkable fact about organic regulation, both developmental and physiological, is that, if the organism is prevented from reaching its norm or "goal" in the ordinary way, it is resourceful and will attain this by a different method. The end rather than the means seems to be the important thing. The significance of such facts for an interpretation of biological organization is obvious.

The maintenance of an organized self-regulating system seems to be a general attribute of protoplasm, but such manifestations of organization as have here been discussed are not by any means a necessary accompaniment of all life. The beautifully coordinated living system sometimes suffers a grievous loss of organization. Tumors, cancers, malformations, and innumerable abnormalities of growth in plants and animals are evidence that the organizing control is sometimes relaxed. Its most radical modification is shown by certain types of cells which may be cultured indefinitely in a nutrient solution and there multiply and grow into shapeless masses of tissue. Such cells remain alive and show certain physiological regulations, since a complete lack of organization would bring death; but they are unable to produce a formed organism where each cell has its particular structure and function, depending on its place in the whole living system.

There are evidently various *levels* of organization, some of which are subordinate to others in a kind of hierarchy. A cell is one such level, and the processes which go on within it maintain a certain independence; but cells are organized into tissues, tissues are grouped into organs, and organs into individual organisms. This organization may be very loose, as in certain lowly plants where most of the cells are alike and the individual can hardly be distinguished from a colony; the mass may be more closely tied together, as in a tree, where there is an indefinite number of leaves and branches but a general pattern for the whole; or it may be very tightly organized, as in most individual animals.

Organization, however different in degree, is primarily a matter of *relations*. Harrison well describes it thus: "Particulate units at any level are not wholly independent of one another. The relations of particles are part of the system and it is their behavior in relation to one another that constitutes 'organization.' . . . No particle or unit can be clearly understood or its behavior predicted unless its reactions with others are taken into consideration."[6]

An understanding of how this organization is set up and maintained is the biological problem to which every

[6] R. G. Harrison, *Cellular Differentiation and Internal Environment*, Publication 14, American Association for the Advancement of Science, p. 77.

other is subordinate and contributory. Whatever repercussions it may have upon other fields of human inquiry, it is thus primarily a task for the student of biology in the broadest sense and must be undertaken on his terms. These terms may have to be enlarged, and we may need to learn the use of new methods of attack upon the problem, but it is life that we are seeking to understand, and life is the province of biology. As Needham warns us, "Organization is not something mystical and inaccessible to scientific attack. . . . It is for us to investigate the nature of this biological organization, not to abandon it to the metaphysicians because the rules of physics do not seem to apply to it."[7]

There have been many attempts to solve the problem of organization. For some biologists this presents no difficulty, and is simply the question of how such a regulatory mechanism has arisen in evolution. During its long course, only those variations which were useful in survival persisted, and through this age-long trial and error the nice adjustments of part and process gradually were developed, by chance favorable mutations, until the present beautifully coordinated organic systems were produced. Surely, these men contend, organization must be something which has thus evolved. That it is not intrinsic in protoplasm is proven by the fact that it is often lost in cases of abnormal growth.

This evolutionary explanation is an obvious one, but it has its difficulties. It can hardly make clear, for ex-

[7] Joseph Needham, *Order and Life*, pp. 7, 17.

ample, how the power of regeneration could have been acquired. There seems little likelihood that all the great variety of injuries and losses which a plant or animal can now repair (including those produced experimentally and which almost certainly would never be suffered in nature) have occurred in its ancestry so frequently that natural selection has had a chance to develop organisms able to repair them. To account for correlative changes such as would be required in the development of a regulatory mechanism has always been a major difficulty for the theory of selection.

Organizing *relations* are easy to observe and measure but are very difficult to explain physiologically. It is much easier to deal with *substances*, and in attempting to understand organization biologists have therefore thought more often in chemical than in physical terms. They have frequently postulated specific formative materials, hopefully expecting that these in some way would translate themselves into organizing relationships. Particularly significant among such are the various growth substances, regulators, and hormones which in recent years have been so intensively studied in plant and animal physiology. Among plants, for example, the effects of auxin have been found to be very numerous and important. It is concerned with cell enlargement, cambial activity, bud inhibition, root formation, leaf fall, and other activities, and thus markedly affects the development of the plant. But it is evident that auxin cannot do all these things by itself. It is the agent, the messenger, by which they are accomplished; but the beautifully coordinated results must come from the presence of just the right amount of auxin, at just the right place, and at just the right time. Something must control the auxin, must act as the headquarters from which the chemical messengers are dispatched. Here is the real problem. "When we discover," says J. S. Haldane, "the existence of an intra-protoplasmic enzyme or other substance on which life depends, we are at the same time faced with the question how this particular substance is present at the right time and place, and reacts to the right amount to fulfill its normal functions."[8] Moreover the secret of the action of such a substance lies not primarily in itself but in the specific organization of the cells upon which it acts. Auxin no more makes roots than a nickel makes a tune in a juke box. It simply sets in motion the activity of an organized system. Not the nickel or the auxin holds the secret, but the structure of the system itself.

The amphibian "organizer" postulated by Spemann is an example of the same difficulty. A bit of the roof of the primitive mouth of the young salamander embryo grafted almost anywhere on the body of another embryo will start a new embryonic axis and thus may make the animal a double one, like a Siamese twin. This bit of living tissue was thought to have in itself important organizing powers; but soon other agencies, simple chemical or physical factors, were found which had essentially the same effect, and Spemann himself finally admitted

[8] J. S. Haldane, *The Philosophic Basis of Biology*, p. 79.

that his "organizer" was but a stimulus, an evocator, and that the real problem of organization lies in the responding system, in the living stuff itself, and not in the trigger which sets this off. Such chemical explanations of organization, despite the enthusiasm with which they have been sought, have not thrown much light upon the problem. Probably not many biologists today would agree with Julian Huxley's optimistic prediction in 1933 that we were then on the verge of reducing the organizing powers of a living thing to a chemical formula and storing it in a bottle.

The beautiful structure of chemical molecules, especially in the proteins with their great size and complexity, has suggested that the form and organization of a living thing may in some way be determined by that of the specific proteins it contains. It is hard, however, to picture a mechanism which would bring this about. Baitsell and others have gone even further and suggested that the organism is itself a gigantic molecule and that the forces which integrate it are the same as those which hold together and organize atoms.

Biophysicists have also offered their explanations. Gurwitsch, impressed by the fact that the fruiting bodies of many fungi, constant and specific in their forms, are produced by a tangled mass of apparently similar fungus threads, believes that a formative "field" exists around the developing structure. Whence this arises and how it operates he is not clear. This general criticism can be made of most field theories proposed by other biologists.

More concrete, however, is the suggestion of Burr and Northrop,[9] who believe that the secret of organization lies in the presence of a characteristic bioelectrical field in and about a living individual, which controls its development. They state the problem in terms of the physics of fields rather than of particles. This is a stimulating idea and well worth developing, but it is difficult to picture exactly how it operates in terms of what we now know about the activities of living things.

Physiological regulation is better understood than that in development and is known to be related to the activity of specific chemical substances. This regulation is extremely delicate, as any one administering insulin well knows, for there is always danger from too much or too little. The normal system, however, controls blood sugar automatically and with beautiful accuracy, an extraordinary accomplishment considering all the things that might go wrong and upset it. We are reminded of Henderson's remark that "sooner or later . . . we come upon the fact that a certain organ or group of cells accomplishes that which is requisite to the preservation of the equilibrium, varying the internal conditions according to the variation of the external conditions, in a manner which we can on no account at present explain."[10]

One of the most spectacular attempts to account for organic regulation has recently come from the engineers. Automatically controlled machines have long been fa-

[9] H. S. Burr and F. S. C. Northrop, "The Electro-dynamic Theory of Life," *Quarterly Review of Biology*, X (1935), 322-33.

[10] L. J. Henderson, *The Order of Nature*, p. 86.

miliar, but their complexity has been raised to an extraordinary degree in the production of the electronic calculator. This is a truly amazing device consisting of thousands of radio tubes connected in a complex fashion by which, almost instantly, huge sums can be manipulated and calculations made which would take a corps of computers years to perform. Such a calculator can store information for later use and thus possesses the rudiments of memory. The principle on which it is built may make possible, its inventors believe, the construction of a machine which will answer abstruse questions and may be said to display some degree of ability to reason. Properly constructed it might even play a moderately good game of chess! Dr. Wiener[11] has shown the marked similarities between the behavior of such a machine and that of the nervous system and believes that the key to a knowledge of the latter lies in the principles developed in these calculators and especially in the so-called "feed-back" mechanism. We must salute those who have built machines which have such fantastic possibilities for the service of man. One may question, however, whether these artifacts really give us more than an instructive analogy with protoplasmic regulation. After all, we are not made of tubes, wires, and gears, but of protein molecules. Our bodies are a triumph of chemical, not mechanical, engineering. The electronic calculator may grow into an accomplished robot, but one doubts if it can have an original idea or write a beautiful sonnet, as protoplasmic systems can.

[11] Norbert Wiener, *Cybernetics*.

We must frankly admit, I think, that, despite our ingenious experiments and speculations, no adequate explanation of biological organization is forthcoming. Despite all the advances in a knowledge of physiology and of the physical and chemical character of living stuff, such a solution seems to be almost as far away as ever. Biology has made enormous strides in the study of processes, of the successive series of chemical changes which go on in protoplasm; but these organizing relations which living things display present a much more formidable problem, and it may be that some new idea, some great generalization comparable to that of relativity for physics, will be necessary before we shall be able to understand the true nature of protoplasmic systems, so deceptively simple to outward view but the seat of that complex organized activity which is life.

The fact of organization has so impressed some biologists that they are even inclined to rank it as one of the basic facts in the universe. Thus L. J. Henderson, a biochemist who thought deeply in these matters, says, "I believe that organization has finally become a category which stands beside those of matter and energy."[12] Needham, in somewhat the same vein, writes: "Organization and Energy are the two fundamental problems which all science has to solve."[13] This is not far from the concept of complementarity proposed by Bohr. The important implications of these ideas are obvious.

[12] L. J. Henderson, *The Order of Nature*, p. 67.
[13] Joseph Needham, *Time: The Refreshing River*, p. 33.

Our problem, though first the task of the biologist, must evidently transcend his domain and enter that of philosophy. The list of philosophers who have undertaken to deal with it is considerable. Most notable among them, perhaps, is Whitehead, who based an important part of his system upon the fact of organization, not only in living things but throughout the universe. Biology for him is the study of the larger organisms and physics that of the smaller ones. The notion of particle he would replace by the notion of organism.

Whatever we may think of these deep matters, it is evident that organization as one sees it in living things is a very real fact, explain it how we will. In any problem dealing with life it must be taken into account. The hypothesis which I wish to propose here is that in the regulatory and organizing processes in protoplasm lies the foundation of what are called the psychological or mental activities in animals and especially in man. From a study of it some interpretations will suggest themselves which may help toward the solution of those great problems which were posed at the beginning of our discussion.

CHAPTER II

BIOLOGICAL ORGANIZATION AND PSYCHOLOGICAL ACTIVITY

We have undertaken the ambitious task of drawing from the resources of the life sciences some ideas useful in the construction of a satisfying life philosophy, and to this end have attempted first to discover what is the most distinctive character of all life, that which most sharply marks it off from lifeless things. What relation is there, you may ask, between the rather technical discussion of this question in the preceding chapter and the profound problems to which we addressed ourselves at the beginning? These were problems about life, to be sure, but life in its uppermost reaches, life expressed not in body but in mind, in purpose, in aspiration, in those manifestations of it which we call the spirit of man. What have the biological facts explored here to do with these higher phenomena of life? Is there anything to be learned from life at its lowest levels which may help toward an understanding of its loftiest ones? We shall find, I think, that there is, and I now propose to explore the possibility that what is called biological organization may indeed be the foundation upon which rest these highest aspects of the life of man.

It is characteristic of living material, as has been shown, that the organisms which it builds grow by orderly progression from one step to the next so that a definite series of bodily structures with specific forms and interrelationships are produced, and that physiological equilibria within them are maintained by constant regulatory adjustments. This progressive, organized, and integrative character of life, its conspicuous and distinctive quality, is most commonly recognized in the development and physiological activity of the body, but it bears a remarkable resemblance to phenomena which are admittedly psychological. There have not been lacking biologists from time to time who, at the risk of being called vitalists and visionaries, have drawn attention to this resemblance between the developmental and the psychological activities of living things, between the facts of growth and those of behavior. The embryologist Spemann, in the last paragraph of his book on embryonic development, says: "Again and again terms have been used (in this book) which point not to physical but psychical analogies. This was meant to be more than a poetical metaphor. It was meant to express my conviction that the suitable reaction of a germ fragment, endowed with the most diverse potencies, in an embryonic 'field,' its behavior in a definite 'situation,' is not a common chemical reaction, but that these processes of development, like all vital processes, are comparable, in the way they are connected, to nothing we know in such a degree as to those vital processes of which we have the most intimate knowledge, viz., the psychical ones. It was to express my opinion that, even laying aside all philosophical conclusions, merely for the interest of exact research, we ought not to miss the chance given to us by our position between the two worlds."[1]

In the same vein the zoologist E. S. Russell writes: "The directiveness of vital processes is shown equally well in the development of the embryo as in our own conscious behaviour. It is this directive activity shown by individual organisms that distinguishes living things from inanimate objects." And again: "The fact is that the common ground of both organic and psychological activity lies in the directiveness or 'drive' which is characteristic of both. We must regard directiveness as an attribute not of mind but of life. . . . Purposive activity, as seen in its highly developed form in the intelligent behaviour of man, is a specialized and elaborated kind of directive activity, concerned mainly with the mastery of his material environment."[2]

Herbert Muller puts the idea very well thus: "'Purpose' is not imported into nature, and need not be puzzled over as a strange or divine something else that gets inside and makes life go; it is no more an added force than mind is something in addition to brain. It is simply implicit in the fact of organization, and it is to be studied rather than admired or 'explained.'"[3]

It is to Ralph Lillie that we owe an especially extensive

[1] H. Spemann, *Embryonic Development and Induction*, p. 371.
[2] E. S. Russell, *The Directiveness of Organic Activities*, pp. 178-79.
[3] Herbert J. Muller, *Science and Criticism*, p. 109.

discussion of this problem. "The general conclusion to which we are led by these considerations," he says, "is that in living organisms physical integration and psychical integration represent two aspects, corresponding to two mutually complementary sets of factors, of one and the same fundamental biological process." And again: "Conscious purpose, as it exists in ourselves, is to be regarded as a highly evolved derivative of a more widely diffused natural condition or property, which we may call 'directiveness.' . . . This psychical integration, so characteristic of the living organism on its conscious side, implies the existence of a parallel physical integration, the two forming together a psychophysical unity. . . . In the characteristic unification of the organism an integrative principle or property is acting which is similar in its essential nature to that of which we are conscious in mental life."[4]

What impresses these thinkers is the striking resemblance between the progressive, regulatory, and (as Russell says) "goal-directed" processes of development and physiological activity, on the one hand, and on the other the phenomenon of purpose, of the drive toward an end, which is the basis of most mental or psychological activities. Life is not aimless, nor are its actions at random. They are regulatory and either maintain a goal already achieved or move toward one which is yet to be realized. A developing embryo, especially if its growth is speeded up by time-lapse photography, certainly *looks* as if it were

[4] Ralph S. Lillie, *The General Biology and Philosophy of Organism*, pp. 50, 196, 200, 201.

moving toward a goal which it is bound to reach in spite of the obstacles which we may put in its way to test the intensity and resourcefulness of its "purpose." Needham's phrase, "the striving of a blastula to grow into a chicken," may be a figure of speech, but to some minds it is not far from actual truth.

In all such questions we are forced to leave the comfortable certainties of the laboratory and plunge into the twilight zone of speculation, where fact and fancy are hard to distinguish from each other but where the shapes of great ideas, tantalizingly vague, move among the shadows. Such speculations are often frowned upon by scientists and the tough-minded generally, and they do lead all too easily to absurdities which tempt one from the path of orderly thinking. Many will smile, I am sure, at the speculations which I shall here propose. But we should lose our fear of being too unorthodox. Adventurous hypotheses to deal with unexplained facts are one of the chief lacks in a world which has piled up more data than it has been able to digest; and speculation, so long as it is intelligent and based on a modicum of established fact, is to be encouraged and may well lead to insights into nature which would not be open to those who never stray from the bounds of orthodox thinking. "We are in danger," says Woodger, "of being overwhelmed by our data and of being unable to deal with the simpler problems first and understand their connexion. The continual heaping up of data is worse than useless if interpretation does not keep pace with it. In biology this is all the more deplorable because it leads us to slur over what is characteristically biological in order to reach hypothetical 'causes.' "[5]

The possibility of a relation between development and purpose—between biology and psychology—is such a speculation. I confess to being attracted by it, for it presents an opportunity of unifying our ideas of life at all levels, and of providing some hopeful answers for at least a few of man's great and ancient philosophical difficulties. I ask the reader's forbearance, therefore, while the argument is pushed a bit further than has usually been done.

The position which I propose to defend—the thesis I am nailing to the cathedral door—is briefly this: that biological organization (concerned with organic development and physiological activity) and psychical activity (concerned with behavior and thus leading to mind) *are fundamentally the same thing.* This may be looked at from the outside, objectively, in the laboratory, as a biological fact; or from the inside, subjectively, as the direct experience of desire or purpose. In the present chapter I should like to explore this position with some thoroughness, and in the final one to consider what implications it may have for the problems which were posed at the start of our discussions.

A conscious purpose and the development of an embryo appear at first sight to be so unlike that a comparison between them seems preposterous. One is an experience in vivid, focal consciousness and is centered in the vast

[5] J. H. Woodger, *Biological Principles*, p. 318.

complexity of the tissues of the brain. The other is certainly not a conscious process in the usual sense of the word and is related to no particular tissue but is an attribute of the entire organism. Both, however, are biological activities, phenomena of life, and to compare them profitably we should therefore first reduce both to their lowest common denominator, protoplasm itself. The activities of nerve cells, the central nervous system, and the brain, structures with which we chiefly associate all forms of psychical activity, may for the moment be disregarded. This is neither the time nor the place to discuss the origin of the nervous system, but good evolutionists can hardly fail to admit that it must have risen to its present highly developed state from very simple beginnings and presumably from unspecialized protoplasm in response to requirements for survival of motile organisms in a complex environment. This conclusion is supported by the fact that the lowest animals and the entire plant kingdom lack a nervous system entirely (or any specialized nervous mechanism) and yet are able to perform, albeit in a sluggish and primitive fashion, most of the activities which in higher forms are under the control of nervous tissue. As Bergson says, we need not assume that consciousness "involves as a necessary condition the presence of a nervous system; the latter has only canalized in definite directions, and brought up to a higher degree of intensity, a rudimentary and vague activity, diffused throughout the mass of the organized substance."[6] To talk about "mind" in a bean plant or a

[6] H. Bergson, *Creative Evolution*, p. 110.

protozoan, or even in a worm, may seem absurd, but it is more defensible than trying to place an arbitrary point on the evolutionary scale where mind, in some mysterious manner, made its appearance. We are dealing here with a quantitative and not a qualitative difference.

All living stuff has definite resemblances to nervous tissue. Protoplasm is irritable, will respond in characteristic ways to stimuli from without. Even a naked mass of it like an amoeba moves slowly about and is continually reacting to the presence of food, water, oxygen, and mechanical factors in a manner quite comparable to that of higher organisms equipped with specialized sensory and motor nerves. Even plants, the most static of living things, will respond to environmental changes by characteristic growth movements or tropisms. The region which receives the stimulus here is often not the one which shows the response, proving that the stimulus can be transferred from one place to the other though not through nerves or other specialized cells. Even more like an animal's reaction is the behavior of Mimosa, the well-known "sensitive plant." The stimulus of mechanical shock at the leaf tip causes the leaflets of this plant progressively to fold together, and finally the whole leaf bends sharply downward. The course of this reaction takes but a few seconds, and a study of its electrical correlates by Burr[7] shows its remarkable similarity to a nervous response.

[7] H. S. Burr, "An Electrometric Study of Mimosa," *Yale Journal of Biology and Medicine*, XV (1943), 823-29.

The effect of past events in modifying present activity, the essential feature of what we call habit and memory in psychological terms, is also evident in organisms without a nervous system. As Jennings has shown, protozoa are teachable and learn to reject harmful substances after a few trials. Even plants can acquire specific rhythms, like those of the "sleep" movements of their leaves, which depend on the duration of light and dark periods, and these "habits" will survive for some time after the external rhythm has ceased.

Mind, of course, involves far more than a reception of stimuli and response to them. What goes on between these two events, in the highly organized nervous system, is obviously of the utmost consequence. Here originate the desires, purposes, emotions, and other aspects of psychical life, together with conceptual thought and the highest developments of intellect. In a famous argument between Jennings and Loeb many years ago, the former showed that a single-celled protozoan (*Paramecium*), far from being at the mercy of its surroundings and responding invariably to their stimulation, behaves very differently depending on its particular physiological state at a given time. Whether, for example, it will swim toward light or not seems to depend in large measure on whether it is hungry or full-fed. In any living thing this organized system, this coordinating mechanism which regulates behavior in conformity with some established standard or goal, this decisive intermediary that stands between sensation and reaction, is the basis of purposive behavior and thus ultimately of mind itself. My argument is that this system is the same as that which coordinates all other vital activities, notably those of development and function.

Superficially the resemblance between the mental and the developmental is close enough so that the case for their identity seems not too implausible. But let us not underestimate the momentous character of such a conclusion. To admit that the developmental norm, standard, or "goal" set up in a growing organism is a manifestation of the same control that guides its behavior, that it is in effect a very primitive purpose, is to grant my whole argument. Concede this, and most important and significant conclusions will follow. Deny it, and the entire hypothesis falls to the ground. This is the decisive step. Here, and not in the upper reaches of the psychical, I believe, must be fought out the major philosophical battle. The issue is primarily a biological one, and unless it is explored as such we shall never finally come to grips with it. It is therefore worth most careful scrutiny.

First, let us examine what is meant by a norm or goal or "purpose" in organic development. It must evidently be a set of conditions or relations, in living substance, which in its operation in space-time results in a progressive series of specific embryonic forms culminating in that of the adult. An understanding of the means by which this is accomplished is at the heart of our problem. The most familiar example of such a developmental norm is the genetic constitution of a cell like the fertilized egg, the living substance of which must have a very specific constitution and organization, for out of it comes a very specific organism. Various suggestions were made in the past as to what this constitution is, but we now know that its most important component is a series of *genes* which control the processes of development. In each race these have a definite distribution among the chromosomes of the nucleus, and in most species the genic constitution of every cell seems to be identical. Sometimes, as we have seen, almost any cell of the body is capable, under favorable conditions, of producing an entire individual. The problem of gene action is being vigorously investigated and with encouraging success; but certainly one of the major enigmas of biology is how these thousands of genetic units, scattered at random throughout the chromosomal complement of every cell, cooperate with such amazing nicety and precision that a complex and highly coordinated individual is produced. Innumerable chemical reactions, accurately timed and located, must be involved. Here the problem of organization is presented most vividly. There must certainly be some sort of controlling mechanism, structure, or system in the egg (and doubtless in every cell) which definitely foreshadows the character of the particular organism into which it will develop or of which it forms a part. It is this organization, whatever it may turn out to be in terms of matter and energy, space and time, which, *as experienced by the organism*, I believe to be the simplest manifestation of what in man has become conscious purpose. Just as the form of the body is immanent in the egg from which it grows, so a purpose, yet to be realized, may be said to be immanent in the cells of the brain.

Physiological regulation shows the same sort of internally organized directiveness toward the maintenance of a specific condition, a steady state. The remarkable processes in homeostasis have already been discussed, by which the various components and conditions of the body are precisely maintained. Any change in them at once calls forth a corresponding increase in an opposing process and thus keeps the steady state in delicate balance. Sherrington has well described these orderly changes which go on within a cell. "We seem to watch battalions of specific catalysts," he says, "like Maxwell's 'demons,' lined up, each waiting, stop-watch in hand, for its moment to play the part assigned to it, a step in one or other great thousand-linked chain process. . . . In the sponge-work of the cell, foci coexist for different operations, so that a hundred, or a thousand different processes go forward at the same time within its confines. The foci wax and wane as they are wanted. . . . The processes going forward in it are cooperatively harmonized. The total system is organized. The various catalysts work as co-ordinately as though each had its own compartment in the honeycomb and its own turn and time. In this great company, along with the stop-watches run dials telling how confreres and their substrates are getting on, so that at zero time each takes its turn. Let that catastrophe befall which is death, and these catalysts become a disorderly mob and pull the very fabric of the cell to pieces. Whereas in life as well as pulling down they build, and build to a plan."[8]

It is this building to a plan which is so characteristic of all life. Such a physiological plan, refined and far more complex in the cells of our nervous system but essentially like the developmental plan, I believe is that which in man can be experienced as conscious purpose. Its roots are deep in the regulatory behavior of protoplasm. Homeostasis is not simply a curious process in physiology. It is the satisfaction of our most basic desires.

It is clearly impossible, of course, to speak of conscious purpose at such a primitive level as this, or even of consciousness at all. From such humble beginnings, however, the consciousness which we experience so vividly must have arisen. In some unexplained fashion there seems to reside in every living thing, though particularly evident in animals, an inner, subjective relation to its bodily organization. This has finally evolved into what is called consciousness. Such an inner relationship is most evident in the sensations experienced when nerves are stimulated, its origin evidently going back to the beginning of the stimulus-response reaction in the simplest of living things. I ask you to consider the possibility that through this same inner relationship the mechanism which guides and controls vital activity toward specific ends, the pattern or tension set up in protoplasm which

[8] Sir Charles Sherrington, *Man on His Nature*, pp. 78, 79.

so sensitively regulates its growth and behavior, can *also* be experienced, and that this is the genesis of desire, purpose, and all other mental activities.

Incidentally, this inner relationship gives us a great advantage in a study of the life of man, an advantage which should be exploited to the utmost. Each of us is *inside* a living organism. The usual method of science is to observe from the outside, objectively. All of our knowledge of plants and animals has come in this way; and, strictly speaking, so has what we know of other human beings. Behaviorism emphasized this laboratory method of studying psychology as all-important. Surely the preferred position which we occupy in our own systems ("between the two worlds," as Spemann says), our ability to *feel* what a complex organism is like, can tell us much about ourselves—and, by inference, about our fellows and even the animals—which could not be discovered in any other way. "The biologist," as Jennings says, "has a more intimate access to a certain sample of his material, for he is himself that sample. Through this fact he discovers certain things about the materials of biological science that he cannot discover by the other method alone. . . . He finds that the things to be studied by the biologist include emotions, sensations, impulses, desires. . . . Thus the biologist has two sets of data, discovered in somewhat different ways, one set being discoverable only through the fact that the biologist is himself a biological specimen."[9] Eddington believes that "consciousness, looking

[9] H. S. Jennings, *The Universe and Life*, pp. 9, 10.

out through a private door, can learn by direct insight an underlying character of the world which physical measurements do not betray."[10] The information thus gained through our own inner experiences is much more vague, however, than that which comes from the laboratory and is far less open to measurement and exact analysis. It depends in no small degree on our particular physiological state at the time it is rendered. Because of this, scientists have always been suspicious of introspection as a reliable basis of knowledge about life and man. For biology to ignore the reports from this inner observer, however, or to deny their importance or even their reality, is to give up any attempt to understand life as a whole. These subjective experiences must ultimately be the concern of the biologist as much as are the actions of genes or the chemical nature of hormones. Biology will need to widen its borders for this purpose and to call into consultation its colleagues in other fields of knowledge, but it cannot disregard such experiences if it is finally to tell us what life really is. In our present speculations, therefore, and particularly in any consideration of desire and purpose, we are justified in gaining whatever knowledge we can from introspection. It may be that the present hypothesis can help to reconcile the conflicting inner and outer aspects of man.

At this point some hard-headed objector will no doubt arise and ask why, if purpose is simply the subjective side of the operation of a regulatory mechanism, we

[10] A. S. Eddington, *The Nature of the Physical World*, p. 91.

should not speak of a thermostat or a gyroscope as having a "purpose" (quite apart from the purpose for which it was made). Certainly one of the new electronic calculators with its vast complexity and beautifully regulatory behavior ought at least to have a purpose, if not a soul, of its own! Indeed, a suggestion not very different from this has recently been made by Norbert Wiener, who looks forward to bigger and better machines which will to all intents display purpose, memory, and some ability to learn and reason. If all this is so, continues our objector, why clutter up the argument with talk of psychical factors? Purpose is obviously the accompaniment and result of a fundamental physico-chemical mechanism, and this mechanism is our only real problem.

He may well be right. One can reduce my argument to absurdity by suggesting that a stone, pushed from the top of a hill, has the "purpose" to roll downward, or that a stretched bow has the "purpose" to shoot an arrow. Any physical system which operates under natural forces may be said to have a purpose to perform whatever changes are by necessity latent in the system. A piece of fireworks has a definite structure which results, when it is set off, in a specific pattern of moving light. Is not a "purpose" to do this present in the pattern of powder and fuses? The step seems not a long one from such a structure to the mechanism of a living cell, vastly more complex but still a mechanism, in which a precise physico-chemical pattern (or "purpose") is bound to unfold in a precise and regulated fashion. Here is involved the whole problem of the nature of a living organism. If this is a mechanism of the sort with which science is familiar, its internal pattern, somehow capable of being felt in experience, can well be interpreted as a purpose. This is the position of all mechanistic philosophy, which regards the psychological event as secondary and merely the result of a physical one. But we should remember that the living mechanism is one of a very special kind and that its organization, and thus the origin of psychological phenomena, may involve principles not yet discovered. An advantage of the hypothesis here presented is that it can be accepted by either the materialist or the idealist as a sound interpretation of purposiveness. It does not take sides. It implies neither mechanism nor teleology, fate nor freedom, but simply attempts to tie together, as identical, the biological and the psychological events. Which philosophical theory—mechanistic, idealistic, or other—will ultimately best explain the facts depends on what the nature of biological organization ultimately turns out to be. Here is where the real issue lies.

But the phenomena of development and of psychology are not quite as far apart as they would appear to be if one limits oneself to the extreme expressions of each. Growth, physiological reaction, and true behavior form an ascending series the steps in which grade into each other imperceptibly, and my argument, I think, is measurably strengthened by this fact.

It has often been said that function is the correlate of

structure. In any living system one cannot separate the processes of growth which lead to the development of the body from those by which the life of the body is maintained. Both are physiological activities, and changes involved in growth are essentially the same as those concerned with the maintenance of vital activities and the repair of tissues. Some are centered in or controlled by nerve cells but many are the activities of other types of cells or of unspecialized protoplasm. It is equally hard to separate some of these physiological processes from true behavior; from activities, in other words, which are commonly held to be psychical in their character. Is breathing, for example, a physiological process or is it a part of the way an animal behaves? The instinctive activity of the simplest organisms seems to be far closer to the physiological than to the mental level. One does not usually speak of instincts in plants, though many of their reactions, such as the tropisms, differ little from the instincts of the simplest animals. Animal instincts range from the simplest sorts of reactions in the lowest groups of invertebrates to the complex and marvelous ones of the bees and ants and reach up toward the essentially rational activities of the higher vertebrates.

The close relation of developmental activities to these instinctive and behavioral ones has been stressed by Bergson and others. "We cannot say," he writes, "as has often been shown, where organization ends and where instinct begins. When the little chick is breaking its shell with a peck of its beak, it is acting by instinct and yet it does but carry on the movement which has borne it through embryonic life."[11]

Lillie makes much the same point. "Development," he says, "has been compared with instinctive activity by many biologists, and instincts have their close affinities with conscious behavior. The development of an egg into the adult animal is a sequence of biological activity which has much in common with such an instinctive performance as the building of a nest; in both cases there is an integrated sequence of morphogenetic or structure-forming activity."[12]

The regulatory character of instinct is obvious. An animal usually tends to act, under normal conditions, in ways that will insure its survival and reproduction. It avoids enemies, captures food, seeks a favorable habitat, and in other ways adjusts its reactions to the environment in a manner favorable to the maintenance of its own life and the perpetuation of its species. In this respect the similarity between these simple types of animal behavior, on the one hand, and biological organization as expressed in development and physiology, on the other, is striking and seems to be fundamental. The primary reason for the rise of higher types of psychological behavior, culminating in mind, seems to have been the necessity of *speedy* regulatory reaction to insure the survival of a motile organism in a complex and changing environment. Here the sensations received are numerous and

[11] H. Bergson, *Creative Evolution*, p. 165.
[12] Ralph S. Lillie, *General Biology and Philosophy of Organism*, p. 172.

varied and the number of possible reactions to them is very great. To determine just which of these reactions should be made to insure survival and a successful life required a series of specialized cells and finally a rather elaborate central nervous system. The more complex problems the species was called upon to meet, the more highly developed this system became, reaching its climax with the brain, active consciousness, and high intellectual ability of man. Consciousness must *mean* something. We shall agree with Henderson, I think, when he says that "consciousness was never produced in the process of evolution merely as an impotent accompaniment of reflex action."[13] William James has put the matter well thus: "Primarily," he says, "and fundamentally, the mental life is for the sake of action of a preservative sort. Secondarily and incidentally it does many other things."[14]

This "action of a preservative sort," it seems to me, is the same kind of action that is found in physiological regulation and the coordinated control of life processes generally. The latter are less obvious and spectacular than the reactions seen in behavior, but both are activities of an organized protoplasmic system. If the primitive purpose to survive is the basis of all psychical behavior, the argument, I think, is sound that such behavior, and thus all mental life, is anchored in the general regulatory activity of living stuff, whether this is in behavior or in development.

[13] L. J. Henderson, *The Order of Nature*, p. 93.
[14] William James, *Psychology*, p. 4.

One might present evidence from many fields in support of this contention, but only a few examples will be offered here.

Mental development in its unfolding often presents a series of steps as precise and predictable as those in embryonic development. Gesell in his classic studies of infant behavior shows that the progressive steps in a small child's psychological development fall into a definite pattern and that these earliest instinctive reactions, the beginnings of the life of his mind, are simply a continuation of the biological (embryonic) activities which brought it into being. The mind has a morphogenesis as well as the body, and both have a definite biological basis. " 'The mind' may be regarded as a living, growing 'structure,' " says Dr. Gesell, "even though it lacks corporeal tangibility. It is a complex, organizing action system which manifests itself in characteristic forms of behavior—in patterns of posture, locomotion, prehension, manipulation, of perception, communication, and social response. The action systems of embryo, fetus, infant and child undergo pattern changes which are so sequential that we may be certain that the patterning process is governed by mechanisms of form regulation—the same mechanisms which are being established by the science of embryology. . . . The growth of tissues, of organs and of behavior is obedient to identical laws of developmental morphology. . . . Already many of the current morphogenetic concepts have more than vague analogy to psychical processes: embryonic field, gradient theory, regional determination, autonomous induction, potency, polarity, symmetry, time correlation, etc."[15] Surely this is strong support for my contention.

In the later part of the organic cycle there are also resemblances between body and mind. It is well known that during its early stages development is much more adaptable than it afterward becomes. Often each cell at its beginning is capable of growing into an entire individual, if isolated. As the organism gets older this capacity to regenerate becomes more and more restricted until in most animals it is lost altogether save for such activities as the renewal of worn-out structures and the healing of wounds. A cell which in the early stages can develop into almost anything suffers a progressive reduction in its potencies until at last there is only one fate in store for it instead of many possible ones. The similarity between such progressive restrictions in developmental potency and the losses in mental plasticity and adaptability with increasing age again suggest a common basis for both.

A similarity between biological and psychological organization exists even in the manner in which they become disorganized. Neither is perfect or infallible. As has been mentioned, the beautifully regulated progress of embryonic development, which usually marches so precisely toward its destined goal, may sometimes be grievously deranged. The disorganization of mental processes resembles that of bodily ones in so many re-

[15] A. Gesell, *Studies in Child Development*, pp. 54, 55.

spects as to suggest that both are the result of disturbances in fundamentally similar mechanisms.

Finally, from quite a different quarter, support for the present hypothesis comes from the concepts of Gestalt psychology. I am not competent to discuss these in detail, but the basic position of this theory is that sensations are not separate and independent things, each nerve cell carrying a specific message which is unrelated to any other message, but that there is a spontaneous and compulsory grouping of them into patterns (Gestalten). These have not arisen from chaotic jumbles of sensation by slowly acquired experience, but are there by virtue of the organizing capacity of the nervous system itself. As Köhler says: "Organization in a sensory field is something which originates as a characteristic achievement of the nervous system"; and again: "The organism is not barren functionally; it is not a box containing conductors each with a separate function; it responds to a situation, first, by dynamical events peculiar to it *as a system* and, then, by behavior which depends upon the results of that dynamical organization and order."[16] In other words the nervous system is morphogenetic. It organizes its chaotic data into forms, patterns, wholes. This is most readily recognized in visual patterns. In a series of apparently meaningless lines an observer will often see particular forms which stand out from the rest. In a puzzle where one is told to "find the elephant," for example, suddenly from the tangle of lines the figure of an elephant stands

[16] W. Köhler, *Gestalt Psychology,* pp. 174, 180.

sharply out. Such patterns are seen as wholes and at once, not as groupings of units. This formative, organizing capacity of nervous tissue which shows itself in the integration of sensations into wholes bears, it seems to me, a striking resemblance to the formative capacity of protoplasm generally, as expressed in development and coordinated maintenance. It certainly is evidence of a basic similarity between the physical and the psychological elements of protoplasmic behavior. Gestalt psychology is one aspect of what may well be called Gestalt biology.

You will perhaps ask what relation there may be between our present hypothesis and the one proposed by several writers, notably Samuel Butler, that heredity and memory are fundamentally the same and even that regeneration is a kind of "remembering." These ideas, especially at the time when they were presented in the challenging form which Darwin's great antagonist knew how to use so well, attracted much attention. Even today *Unconscious Memory* is good reading. But the great advances in biology since Butler's day, and especially the development of modern genetics, make it clear that he was mistaken. It seems certain today that what is inherited is a series of genes, well insulated from external change, and that these do not accumulate a record of the bodily alterations acquired by the organism during its life in the way that its nerve cells acquire a memory of past events or a habit which these induced. There is a resemblance between Butler's ideas and those here presented only in that both regard biological and psychological facts as having a common basis. Our hypothesis sees in purpose, however, rather than in memory, the process common to life and mind.

One might well extend the argument further, but perhaps enough evidence has been presented to convince the reader that the simplest types of psychological activity bear so many resemblances to the organizing processes of protoplasm, as manifest in the regulatory character of all vital activities, that it is at least reasonable to regard them as having a common basis and origin. But many, I am sure, are ready to remind me that it is not tropisms and instincts and the behavior of plants and the lower animals which chiefly interest us. Man is the only organism we wish to explore; man, the flower and summit of the evolutionary process; man, the possessor of intelligence and rational thought, and subject, therefore, to those doubts about his nature and destiny which have so troubled his kind from the beginning. The arguments here presented are of significance, perhaps, for biologists and genetic psychologists, but what do they tell us, one may ask, about the upper reaches of man's mind and those higher qualities of his spirit which set him up above the brutes? What relation does biological organization have to this vividly conscious, thoughtful life of ours?

The only answer I can offer is the old one of upward evolutionary change. Most will concede that man's mind as well as his body has ascended to its present high estate from humble origins in lower forms of life. Reason and abstract thought, the possession chiefly now of man alone, have evidently come from the simpler psychical life of his ancestors. The steps through which this has been accomplished one can guess by observing the behavior of the higher vertebrates and particularly the members of the primate stock. Perhaps, as Bergson believes, the development of man's intellect and capacity to reason has resulted from his use of tools, a trait not shared by other animals. This has enabled him to see relationships among objects outside his own body and to recognize uniformities in nature and the existence of causes and their effects. Out of this has grown the ability to reason. The power of imagination endows him with the capacity of abstract thought which has proven so vastly important to his progress. One cannot minimize the tremendous importance of this development of man's mind to himself and to the world, for the advantage it gave him has set him above the rest of animate nature far more than his very modest bodily attainments would ever have done. It is hard to see in it anything qualitatively new, however. The argument still stands up that intellect is simply a very complex expression of the regulatory character of all protoplasmic activity. Secondarily, as James says in the quotation of a moment ago, the mind does many other things. These are of the greatest significance for our lives, but from the viewpoint of evolution they are indeed secondary to its major biological function.

When we come to the origin of man's spiritual qualities—his love of beauty, his aspirations to virtue and godliness, his yearning for understanding—qualities which have been the hope and despair of his race, we are confronted by a more difficult question. Whence have such traits arisen? They seem, indeed, to be something new among living things. Most of them are either unknown among the lower animals or represented there only in rudimentary fashion. They are surely most significant and worthy of study, and evidently tell much about the complex biological systems which we are. These I shall discuss more fully in the concluding chapter.

Finally, another serious objection to my whole argument will occur to many. The goal of biological development is a single complete organism. Its attainment and maintenance in a given state fulfill the primitive "purpose" in the egg. What relation can there be between such a single, persistent goal and the stream of thoughts and purposes which make up man's rich conscious life? If mind indeed is at bottom a glorified expression of the organizing power of protoplasm, how can we explain its constantly changing content, its vast versatility?

This is not an easy question, but I believe it can be answered. The stream of consciousness, to be sure, is far more complex than the plodding, single-track purposiveness of development. Nevertheless, the latter process is not as limited as it often appears. A single organism is its goal, or a specific functional state, and the persistent drive to attain this goal despite all obstacles is the most significant thing about life. The exact character of the organism produced, however, will depend, in some measure at least, on the environment in which it develops.

There is a variety of primrose, for example, which will produce white flowers if grown in a warm greenhouse, but red ones in a cooler one. Here the end, the goal, the "purpose," is quite different depending on the temperature at which development proceeds. The small vinegar fly, *Drosophila*, which is such an important animal for modern genetics, can be grown under experimental conditions as to temperature, food, and other environmental factors quite different from those met in normal development; and when this happens, individuals may be produced which are markedly unlike the typical flies. The literature of genetics is full of such examples.

In all these cases the genetic constitution of the organism is not changed, but the way in which this expresses itself in development is very different depending on the conditions under which development takes place. Genetics has learned that most genes are not ones "for" certain characters but that what each will do depends on the internal and external conditions under which the gene operates. It has not one role but a whole repertoire. Physiological equilibria, delicately maintained under one environment, may also change their levels as conditions change.

The relation of these changing developmental goals to a stream of conscious thoughts is not too remote, for psychological goals also change as the environment changes. A simple animal like a mollusk, anchored in one spot or moving but slowly, is adjusted to a relatively stable environment. When this is altered, it changes accordingly, but the number of necessary alterations is slight. Its "purposes" are few and constant. Its "mental" life must be sluggish in the extreme! An insect, moving about much more rapidly and exposed to a far more varied environment, is continually adjusting its inner state to these changes and has a far more complex psychical activity, as its well-developed nervous system indicates. The mental life of man, however, has been stepped up to so much higher a level that it seems to be quite different in character from that of animals. His environment is vastly more complex than theirs not only because of the development of his intellect and social relationships but because he responds to symbols—pictures and the spoken and written word. Among civilized men the number of symbols and the reactions they induce—the ideas they convey—are exceedingly numerous and for many persons are among the most important factors in the environment. Beyond all this, man possesses an ability which further enriches the life of his mind and sets him sharply apart from the brutes: he can conjure up before him, so to speak, images of events in the past or in the possible future, hypothetical conditions, assumed relations. Thus in the quiet of his chamber, physically at rest and subject to no change in his surroundings, he can indulge in that highest of intellectual activities, abstract thought. In the mind of an artist there may arise the idea for a great picture, and without this idea any collection of paint and canvas is meaningless. These are among those secondary qualities of the mental life of which James speaks. Memory, imagination, and abstract thought, the ability to bind the past and future into the present and to single out and manipulate particular qualities or relations in the environment, these are great accomplishments and enrich the mental life of man far above that enjoyed by any other animal. How this is accomplished psychologists cannot tell us. How we are able to say, lo, I will close my eyes and fill my mind with the sensations and actions and experiences of yesterday, or with a picture of events which I should like to bring to pass tomorrow—this is still a mystery. In the terms of our present thesis, however, the mystery is one of protoplasmic mechanics, for each of these ideas, these thoughts, these mental pictures is ultimately resolvable, I believe, so far as its *mechanism* is concerned, into the same sort of an organized norm or goal as is found in the protoplasmic system which controls and regulates development. Ideas were presumably desires or purposes at first and expressed themselves directly in action; but such psychological events do not *necessarily* lead to action. A continuous stream of them may fill the mind without producing any visible physical changes in the body. The wish may thus indeed be father to the thought.

Interpreted in terms of the present hypothesis, therefore, the whole conscious life of man, rich in ideas, in inspirations, in intellectual subtleties, in imagination and emotion, is simply the manifestation of an organized biological system raised to its loftiest levels. Upon this the outer world impinges as a series of sensations, real or imagined, and out of it come actions, either actual physical responses or the more subtle ones of the mind. What takes place between these events is, at bottom, the regulatory activity of the protoplasmic system. In its lowliest expression this appears as regulatory control of growth and function. This merges imperceptibly into instinct, and from these simplest of psychical phenomena gradually emerge the complex mental activities of the higher animals and finally the enormously rich and varied life of the mind and spirit of man. At no point is there a sudden break, a radical innovation. The complex has come from the simple by a gradual process of evolutionary progression. The basic phenomenon from which all this ultimately arises, the fact that living things are organized systems, is the fundamental problem, still unanswered. Upon its solution will depend our understanding not only of biology and psychology but of the whole of man.

Such is the statement of my hypothesis. There is nothing very novel about it, for ideas of this sort have often arisen in the minds of biologists and philosophers, though no one, perhaps, has pushed them to quite the extremes that I propose to do. They are highly speculative and will seem nebulous and of little meaning to many hard-headed scientists. They suffer, too, from a certain psychological naïveté, for one can hardly hope to explain the vast complexity of man's mind in such short space without oversimplifying the problem. We have set out, however, to try to bring the facts of biology and psychology under a single aegis, and I believe that the hypothesis here presented offers a defensible attempt. These ideas would be interesting and worth pursuing for their own sake and their relation to the sciences of life and mind; but their chief importance, it seems to me, is in their bearing on some of the great problems of man's nature and his relation to the universe, problems which have troubled him from the beginning and which were set before us at the start of these discussions. The significance of such a concept of life and mind for the solution of these problems I shall discuss, albeit most inadequately, in the final chapter.

CHAPTER III

SOME IMPLICATIONS FOR PHILOSOPHY

In the first chapter the conclusion was reached that the fact of *organized* growth and activity, leading to the production and maintenance of those self-regulating structures so appropriately called organisms, is the most distinctive feature of living things. In the second, your attention was called to the close resemblance between this biological organization, on the one hand, and desire, purpose, and the higher phenomena of mental life, on the other. I then tried to persuade you that these two are manifestations of the same underlying fact, whether observed from the outside by a biologist studying an organism or experienced from the inside by the organism itself.

Such a hypothesis may be interesting as a philosophical speculation, one will say, but what can it do to answer the questions we asked ourselves at the beginning of these discussions? Can it lead to any new insight into the nature of life, especially as expressed in the remarkable species to which we belong, and into our relationship with the rest of the universe? I believe that it can. My concluding task is to present reasons for this belief and to discuss six such problems upon which the hypothesis here proposed may throw some light.

The first of these is one which has tormented man throughout his history: the relationship between those two parts of him which seem so vastly different—his body and his mind. These evidently have much to do with each other, but just how are they related? Is the body—tough, tangible, and material—the part of a man which is truly real and the mind but a curious result of physical forces only, an epiphenomenon, something that rides along on the crest of the material wave but has no control over it, no existence independent of it? Such has been the belief of many men through the ages, and such is the creed of scientific orthodoxy today, confident in matter, suspicious of all else. Or is mind, with those deeper feelings which accompany it, the essential member of the pair, autonomous, ruling matter, and in some mysterious way the true and permanent reality, and all else illusion? Yes, say the poets, the dreamers, the believers, those who walk by faith, not by sight. Whichever is dominant, this vexing dualism, splitting man in two, has long wrought confusion in his thinking. But if the view of life which I have presented is correct, such dualism is apparent only, and not real. Body and mind are simply two aspects of the same biological phenomenon. The first is no more real than the second, for they are one. The pulling together of matter into an organized living system is what we feel as a mental experience.

The advantage of this point of view is that it does not

commit us to the position of either the materialist or the idealist. It simply asserts that there is one basic problem, common to both, which must be solved before we can understand either. The materialist may well maintain that this view is a sound one and essentially what he has always believed. Since matter and energy in his opinion are the only true realities in the universe, to assert that what we know as mind is nothing but the organizing process in the development of those material systems which living things are, is to admit the truth of his position that mind is simply one expression of the activity of such a system. The profound effects of drugs upon mental phenomena of all sorts, the clear relation between specific regions of the brain and psychical states, and the notable achievements of neurophysiology generally all lend support to such a view.

But this hypothesis, if carried to its logical conclusion, makes such severe demands that the idealist may fairly entertain a doubt as to whether matter and energy, in the traditional sense, can satisfy them. If material changes related to the processes of growth and function account for mental changes as well, and if the simpler desires and purposes are thus rooted in the regulatory character of protoplasm, we cannot restrict the responsibility of this material system but must expect it also to explain the highest qualities and triumphs of the mind. These are as much an expression of protoplasmic integration as the form of a leaf or the growth of an embryo. The organizing capacity of living stuff so clearly manifest at the lower biological levels is equally effective far above them and is the source of new ideas, of high aspirations, of lofty flights of the creative imagination; the means, indeed, by which man launches out into the deep and challenges the unknown universe. Surely matter and energy which can thus come to flower must be more than the simple things which the student of the physical sciences sometimes considers them. Indeed, our respect for the complexities and possibilities of the material world has vastly increased during the past generation, largely because of the labors of some of these physicists. The possibility emerges that instead of matter and energy explaining life, life—as a very special category of the physical universe—may in time make contributions of its own to our knowledge of matter and energy. Genetics and physiology have thus already posed new problems for the physical sciences. Needham remarks that "empirical discoveries on the purely biological level thus serve as stimuli to the physiologist to investigate processes which his methods alone would never have revealed in the first place."[1] J. S. Haldane goes even further: "As the conception of organism," he writes, "is a higher and more concrete conception than that of matter and energy, science must ultimately aim at gradually interpreting the physical world of matter and energy in terms of the biological conception of organism."[2] The English mathematician and philosopher J. W. N. Sullivan agrees. Says

[1] Joseph Needham, *Order and Life*, p. 22.
[2] J. S. Haldane, *Mechanism, Life and Personality*, p. 98.

he: "It is possible that our outlook on the physical universe will again undergo a profound change. This change will come about through the development of biology. If biology finds it absolutely necessary, for the description of living things, to develop new concepts of its own, then the present outlook on 'inorganic nature' will also be profoundly affected. . . . The notions of physics will have to be enriched, and this enrichment will come from biology."[3]

If this is true, is it not possible that, in like manner, a study of those qualities of the human spirit which we regard as the highest expressions of life may throw light upon problems which neither biology nor the physical sciences alone could ever solve?

Such speculations are perhaps of little merit now, but in this ancient problem of the mind and the body it is not without value to remember, whatever our views about them may be, that they have a common denominator, a fundamental similarity, and that whatever commerce there is between them is based upon this fact. A drug affects a mental process, or an emotion a physiological one, not directly and specifically but through its influence upon the entire organized system. The rise of psychosomatic medicine is a recognition of the fact that the entire individual, mind and body, is the important entity in health and disease, and that no one part can suffer or be ministered to without affecting all the rest. The patient is a biological system, not a collection of organs

[3] J. W. N. Sullivan, *The Limitations of Science*, pp. 188, 189.

and symptoms. This system is the unity, the synthesis of the mental and the material, because both are aspects of it, and an understanding of what such a living system is and how it works will ultimately solve the ancient enigma which we have been discussing.

The second problem for an understanding of which the hypothesis here presented may be useful is that of motivation, of the desires and purposes which drive us on and which are the foundation of all mental life. We have discussed the similarity between the regulatory character of development and physiological activity, on the one hand, and that of behavior, instinct, and conscious purpose, on the other. A standard, norm, or goal set up in living stuff but still to be reached creates the desire for its attainment; and if this attainment is prevented, or if the equilibrium is thrown out of balance, the organism experiences unease, pain, or distress of body or mind.

The desire to reach such a goal can hardly be separated, at the lower levels, from the purpose to do so. Around this idea of purpose there has long raged one of those philosophical polemics which touch a problem so fundamental and so difficult that it seems to defy solution. Aristotle distinguished *efficient* causes, which produce effects through direct and evident physical means, from *final* ones, which are goals or purposes in the mind and effective through man's moving toward them. Final causes, as it were, leap across a gap to accomplish their effect and seem not to require the operation of any mechanism. It is the existence of this type of cause which science often has so strenuously denied, since it seems to involve a mysterious non-mechanical, non-material agency. If I wish a cup of coffee and pour it out, the real cause of this action is not the purpose which exists in my mind, says the psychologist, but a complex chain of physical and chemical steps in my nervous system. Purpose is often a "fighting" word nowadays. Whitehead remarks rather quizzically that scientists who are animated by the purpose of proving that they are purposeless are an interesting subject for study. An advantage of the hypothesis which I have here been defending is that it eliminates the antithesis between efficient and final causes since what appears in the mind as a purpose, later to be realized in action, is *the same thing* as the physiological, protoplasmic norm or "goal" set up in the brain and coming to realization through a series of regulatory processes.

Such a hypothesis, it seems to me, if it is sound, would dispose of the difficulties involved in the concept of final causes and teleology generally. This has long been the bogey of the scientist, particularly of the biologist, since it is often carelessly invoked to explain natural phenomena. The plant is sometimes said to grow toward the light "for the purpose" of illuminating its leaves, or because the plant "needs" light to make food. It is clear that to call such behavior purposeful and to fail to point out to the student the physiological mechanisms involved and the long evolutionary history that lies behind them is vastly to oversimplify the problem in his mind and to give him a false idea of the character of living organisms.

The teaching of elementary biology is too often vitiated by such naïve comparisons. The present hypothesis, although it contends that these reactions in their essential nature can be regarded fundamentally as manifestations of purpose, recognizes this purpose to be at a far lower level than that with which the student is familiar in his own mind.

Much of the difficulty in accepting the idea of purposeful action by plants and animals comes from the assumption that this implies an ability to do what is best for them, what will always tend toward their survival. This would indicate an unexplainable ability of living substance invariably to do the right thing, to adapt itself in a favorable manner to changing conditions. This it seems by no means to possess. Many of the races of plants and animals which have appeared as mutations certainly lack it. The type of corn known as "lazy," for example, grows flat on the ground, and no amount of propping up will induce it to become erect. It is not weak-stemmed. It is naturally prostrate, held there by the unusual distribution of its growth substances, in turn controlled by its genetic constitution. In our sense of the word it has a purpose—to grow horizontally—but this would soon lead to its extinction in nature. Organisms often fail to act in such a way as to favor their survival. As Jennings has said, to make mistakes is one of the characteristic phenomena of biology. In other words, we can and often do have purposive organization without adaptation. Each individual has its own genetic equilibria or physiological norms, its own primitive purposes, and new ones which arise may well be unfavorable. Natural selection eliminates these and preserves individuals which tend to react in a favorable way, which have "purposes" that are conducive to successful life and survival, which "want" the right things. There is really more here, I believe, than a figure of speech. We are justified in saying that the ultimate interpretation of why a leaf turns toward the light is that such action is a regulatory and thus a purposive but not necessarily an adaptive phenomenon. This does not mean, of course, that the mechanisms by which this is accomplished—hormone distribution and other physical changes—are not essential. These are the means by which a specific regulatory behavior is attained. Such behavior enables the plant to survive, however, by virtue of long evolutionary selection and not because of any innate tendency to react in a favorable manner. Adaptation, the beautiful and precise fashion in which organisms so commonly respond to their surroundings in such a way as to live and reproduce successfully, is simply the result of past competition between many different inherited individual purposes.

This vexing problem of purpose is somewhat simplified, I think, if we thus are able, in the individual organism, to reconcile efficient with final causes, mechanism with teleology, by showing that they are essentially the same. In all this, again, the nub of the problem, the question on which everything else depends, is the nature of this organizing, regulatory behavior which runs through all life and seems always purposive, making action conform to a goal set up in the system. The ultimate explanation advanced for this basic biological fact will determine our interpretation of purpose at every level from the growth of an embryo to the aspiration of a saint.

The third question is related to this—the ancient one of *value*. Why do some things seem inherently desirable and worthy of our devotion, and others unattractive or abhorrent? The elusive quality called beauty, for example, surely exists and for many is a supreme value in the universe. But what is it, why do we admire it, and how can it be recognized? It bears no sure sign upon it, nor is there any yardstick to measure its degree. Men often differ as to its presence, as to whether a painting or a piece of music is beautiful or not, and the tides of opinion rise and fall from one generation to the next. So is it with what is called the right; all are far from agreement as to what acts and attitudes are right and what are wrong, and moral codes often change at frontiers and with time. Yet the right is surely another of the supreme values, and men innumerable have died for it. A chief cause of confusion in the world today is that men cannot agree as to what values are, cannot establish for them any code which will command wide acceptance. And yet, uncertain as values often seem to be, there is something in them which demands our allegiance and utters an imperative that we cannot disregard.

This problem comes down at last, I think, to the direction and character of the goals of life. Those desires set up in protoplasm and determining the ends toward which an organism moves are not random ones. Many lower animals show preferences among the various tastes, odors, and colors of which their sense organs give report. The growing parts of plants move toward or away from light or the earth's center. Even the naked protoplasm of a slime mold pushes out toward certain objects and pulls back from others. Of course, these reactions have long been winnowed out by selection and are usually such as favor the survival of the organism; but, if our hypothesis is sound, they are the primitive beginnings of what in man have become conscious preferences, judgments as to value. The fact that these preferences follow certain general directions, vague though these often are, is certainly not without significance for the problem of value. We may not all agree as to the beauty of a given object, but the fact that training can increase an appreciation of the beautiful and that the accumulating opinions of men through the years seem to move toward ever greater agreement encourages us to believe that human judgments as to values have some meaning. We may well gain insight on these problems by studying the preferences set up in living things, the course and orientation established by those protoplasmic systems we have been considering. They are vanes which show the way the winds of the universe are blowing. If there is any harmony between our instinctive preferences as living things and any standards of value established in nature, its basis lies in this organized stuff of life.

The fourth question on which our hypothesis may be of help grows directly from the last two. It is another ancient battleground—the problem of free will versus determinism. Are we really free to move toward the fulfillment of our desires, toward the accomplishment of our purposes, or is this sense of freedom an illusion and are we actually machines, as truly determined and as subject to mechanical laws as an automaton? Perhaps nothing new can be said on this venerable issue, but with the progress of knowledge it continues to be stated in fresh terms.

Evidently *something* must determine our acts, or else we must believe in chaos. The issue is whether *we* determine them or whether something *not* ourselves does so, something alien to us, either within our bodies or without. Surely we have the feeling, and so strongly that it is taken for granted as the basis of our moral and social codes, that we do determine our own deeds, that we are responsible for them. To be sure, we are often buffeted by our environment, often kept from our heart's desire; but in normal individuals this does not shake the feeling that we are still the masters of our fates, the captains of our souls.

And yet almost the whole weight of modern physical science speaks in no uncertain terms against this conclusion. What are our desires, that they should turn aside the inevitability of a chemical reaction? How can we expect to modify so truly sublime a certainty as the Second Law of Thermodynamics? To do so, tough-minded men tell us, is to give up the battle, to believe in fairies, to expect miracles, to turn our backs on the whole faith and fabric of science. This dilemma between what we feel is, and what we think must be, has troubled generations of men and is perhaps no nearer a solution today than ever. Says Henderson: "It is a strange irony that the principles of science should seem to deny the necessary conviction of common sense."[4] The hypothesis which has been presented here may help to frame the issue in a somewhat different and perhaps a clearer light.

I have interpreted a purpose as essentially similar to a developmental or physiological equilibrium, a tension or prospective goal set up within the living system and normally realized in action if external forces do not prevent. As an open-field runner in football side-steps one tackler, dashes ahead, dodges another, perhaps to cross the goal line, perhaps to be stopped short of it, but always with the transcendent purpose in his mind of making a touchdown, so in a far less dramatic and obvious fashion an embryo moves toward its "purposed" goal, surmounting as best it can the obstacles in its way. Such obstacles, to it and to us, are often too powerful and prevent the attainment of an end. But this sort of bondage we recognize as such. It does not destroy our belief that we are really free, for it does not affect the fundamental mechanism through which purpose is translated into deed. This translation is clearly part of the same dynamic system as the setting up of the purpose itself. Furthermore—and

[4] L. J. Henderson, *The Order of Nature*, p. 92.

this is the important thing—the ego, the self which does the purposing and acting, is also a part of this same system, is indeed the core of it and the sum of all the organizing relations of the individual. The purpose and the purposer are one. It is we who will and do, not some agent foreign to us. Surely, if this is so, we do what we will, for the desires that arise in us are an essential part of us, and to speak of compulsion here seems foolish. How these purposes and how we ourselves arise through the organizing activity of living matter is the real question. If this is all part of a rigid, determined, material order, then our sense of freedom, which is merely the coincidence between ourselves and our acts, is, in the strict sense, illusion, as so many philosophers have maintained; but it is an illusion which bears the stamp of reality and for practical purposes *is* real.

But there is another way of looking at this problem. Life is not static. It is creative. It begets novelties. Each organism is a new and different event. It may be that this creativeness is expressed not only in genetic changes in protoplasm but in new purposes and ideas arising within us, and that the organized system we have been discussing in some way calls these up, brings them to pass independently of external compulsion, and therefore that in a more distinctive sense it is a free, creative agent. As Whitehead says, the psychical is a part of the creative advance into novelty. Vitalism, you will say; but the inner springs of creativeness in protoplasm still are so obscure that we should not be dogmatic about them. The human spirit, like an explorer in an unknown sea, may really be steering its course as it will, free and untrammelled, with nothing to guide it but its own inner directive. Whence this arises may be one of the sources of the new and unpredictable in the universe.

The real issue in the problem of freedom is again the fundamental character of biological organization, of what it is that sets up self-regulating mechanisms in a living thing, and ultimately in the nervous system of man, and which thus creates purposes and fulfills them in action. To bring into this ancient controversy the conception that man is such an organized system, with all that we have seen this to imply, makes possible, it seems to me, a somewhat clearer statement of the issue involved.

A fifth question which our hypothesis may tend to clarify and which has already been mentioned is that of the self, the ego, the *individuality* of a human being. One of the most noteworthy features of living matter is that it not only tends to pull itself together in an organized fashion but that the systems it thus creates form separate and distinct organisms. Almost of necessity this is so. The goal of the organizing process seems always to be a *single, whole individual*. This is evident not only in normal development but especially in cases of regeneration, for if a part of an individual is removed the course of growth will be altered in such a way that the missing part or an equivalent structure is replaced and a whole organism thus restored. There is reason to believe that every cell, at least in its early stages, is capable, if separated from the rest, of growing into an entire individual. A single whole is immanent in all its parts. When a protoplasmic system is continuous, it tends to produce and maintain a single organism. Where parts of it are separated from each other, either actually or functionally, each of them normally does this. Protoplasm always comes in separate packages. One of the most fascinating aspects of the study of organic development is to observe how universal is this tendency to form single organisms. Such organisms may be combined into societies, or may actually be united into colonial masses, and in some, as in most plants, the individual is a rather loosely knit aggregation of multiple parts. Nevertheless the product of the developmental process is almost invariably a single, functioning, coordinated, living unit. The significant fact in all this is that the individual is the sum of all the organizing relations within it. It is the center of integration, the core of the processes of coordination, all of which are knit together into a single physical and psychological whole. The result of all this, of course, is the production of countless centers of organization, each of which thus has its own psychical identity. These selves in the lower organisms are relatively simple, but in the upper portion of the evolutionary series among the animals they have developed into the complex beings which culminate in human personalities.

Such a living individual is a remarkable thing. It is a unit not only in space but in time. It persists. Its history is a continuous progress, not a mere repetition of the same reactions. In the pulse and stir of time it maintains its own identity. Bergson has emphasized this fact. "Like the universe as a whole," he says, "like each conscious being taken separately, the organism which lives is a thing that *endures.* Its past, in its entirety, is prolonged into its present, and abides there, actual and acting. How otherwise could we understand that it passes through distinct and well-marked phases, that it changes its age—in short, that it has a history?"[5] However long its history may be, however varied its surroundings and its activities, it remains the same individual. Matter enters and leaves it, and its material constitution may be replaced many times, but its fundamental organization is unaltered. It is unique; not just one of a long series of similar units, but unlike—or so it seems—any other individual that ever lived. An unchanging genetic constitution is doubtless of basic importance here, but characteristics acquired during the individual's history—bodily skills, memories, tastes, and prejudices—are also built into the persisting self. For any living machine to maintain the delicate physiological balance necessary for life is remarkable enough, but to preserve its specific character, unaltered by the flux of chemical and physical change, is indeed past our understanding now. Human personality, tenuous as it may sometimes seem to be, is of surprisingly tough fiber. The knot of norms, goals, steady states, potencies, and purposes of which it is composed is almost impossible to loosen. To kill it is easy, and to direct the

[5] H. Bergson, *Creative Evolution*, p. 15.

course of its development not difficult; but to break it down and make it into something different, as a sculptor does with his clay; to shake it free from its past, to destroy its identity—this the organized pattern of personality most successfully resists.

Around the fact of individuality centers a whole group of problems. What is human personality? Has it a value of itself, or is this submerged in the greater value of a society? What is this elusive thing we call the soul? What prospect is there that it may be so sturdy as to survive the disorganization of the body? Can it communicate directly with other souls? Here also arise those great ethical questions of the relationships between human individuals, of selfishness, hatred, altruism, and love. These are questions of the utmost importance not only for an understanding of the universe but for their bearing on the philosophies of men. The hypothesis which has been presented in these pages will certainly not solve them, but it does suggest that they all necessarily arise from the one fundamental fact which underlies the argument here. Whatever one may think of psyches, egos, or souls, they all originate in this remarkable process by which living matter pulls itself together into integrated and organized self-regulating systems. The real problem, as has been pointed out repeatedly, is the character of these systems. The soul is the internally experienced aspect of bodily organization.

Again the materialist can readily accept our conclusion. The soul to him is *simply* the psychical aspect of the material bodily system and nothing more. The ties between physiology and psychology are still obscure, but the latter, he maintains, must certainly be subordinate to the former. The idealist may not be convinced, however. This fact of the organization of living stuff is the very nub of all these problems and until it is disposed of, says he, no certain conclusions can be drawn. Organization, as some philosophers think, may be one of the major categories and may control, rather than arise from, matter. Perhaps these centers of organization are primary things, not secondary ones. Perhaps they may even exist independently of the matter in which they are now embodied. Perhaps, as Schrödinger suggests,[6] they are each a part of a universal spiritual whole. All these ideas the materialist will look upon with amusement or indignation, depending on the toughness of his mind; and, indeed, in the clarity of the laboratory it does savor of the preposterous to think of the soul as anything but a temporary phenomenon, dependent on a series of complex chemical reactions in the body. But the universe is a remarkable and often unpredictable place, and unexpected things keep coming out of it. The obvious is not always the true, as physics has learned so well. Living matter may conceal mysteries deeper and more difficult to comprehend even than this. To say that the soul has nothing to do with the body is foolish. To say that it is a mere accessory to the body may prove equally so. All we are maintaining here is that soul *and* body are manifestations

[6] E. Schrödinger, *What Is Life?*, pp. 88-91.

of the same basic phenomenon, are a fundamental unity. Whence this comes and whither it may go is the ultimate question. Meanwhile the fact that living nature so invariably expresses itself in individuals should emphasize their importance in the nature of things, and the fact that the highest manifestations of the life of man come not through groups but through single human personalities should convince us that these are worthy of our deepest respect and concern.

Out of this problem of the origin of the self and human personality grows that deep final question of what man's nature really is and what is his place and significance in the universe. To seek from our hypothesis aid in the solution of these questions may seem to exalt the fact of biological organization far beyond its natural limitations; but even here I believe that we may gain from it some suggestions which are not without value.

Man is indeed the paragon of animals. Arisen in a few score millennia from the rank of a second-class mammal by his mastery of the power to reason, he has gained ascendancy over every living thing and is set off from the rest by differences which mark him as unique in all creation, the crown and climax of the evolutionary drama. But a human being, the organized self in which the life of man is expressed, is far more complex than a study of his evolutionary history might lead one to expect. He is no mere glorified robot, ruthless, weighing everything in the scales of survival and physical satisfaction. He is a vast deal more than a bundle of purposes with an intellect to help accomplish them. From far down within him, in that deep subconscious matrix where matter and energy and life are so inextricably mixed together, there surge up into consciousness a throng of emotions, longings, loves and hates, imaginings and aspirations, some exalted and some base, which form the most important part of what he is. Here are not only the passions, lusts, and cravings of a species but newly risen above the level of the beasts, but qualities foreign to the brute creation, longings for higher things than they can ever know. "Man," says Du Noüy, "is not merely a combination of appetites, instincts, passions and curiosity. Something more is needed to explain great human deeds, virtues, sacrifices, martyrdom."[7]

Man is stirred by the marvel of beauty in the world around him. Imagination is his alone, the capacity to build in the chambers of his mind things never seen before and thus to create the arts and the sciences. For him the noble and the good exist, and he aspires to reach them. Love for his fellows and a desire unselfishly to serve them have reversed the jungle code and given him a vision of the brotherhood of man, whether he calls it the Paradise of the Proletariat or the Kingdom of Heaven. He is consumed with eagerness to learn the truth, to penetrate the secrets of nature and thus to push forward the frontiers of human knowledge. The wonder and mystery of the universe overcome him, and he falls on his knees in reverence. The conviction stirs in his heart

[7] Lecomte du Noüy, *The Road to Reason*, p. 234.

that he is not alone in it but that something not unlike himself is there with which he can hold communion.

These qualities are far from universal in our race, but the goals which they erect, however differently described and varying in detail, have been acknowledged by all sorts and conditions of men as the highest expressions of human life. These inner urgencies, these passionate and imaginative longings for something higher than he yet has found, are the expression of man's spirit. This is a great and mysterious thing. It is no minor or accessory part of him but essential for his very life. Its values are the highest that he knows. The universe comes to flower not in atoms or galaxies but in poets and philosophers, in scientists and saints.

To separate these yearnings of man's spirit from the lowlier desires by which he gains his food and propagates his race is hardly possible. Their germs are stirring in the beasts. Maternal affection, herd loyalty, cooperative effort, faithfulness—these can be found in many animals, and what we know of human evolution strongly suggests that man's nobler qualities, which now seem so distinctively his own, slowly emerged as he rose to his present high estate. The theme of my argument has been that a continuous progression exists from the biological goals operative in the development and behavior of a living organism to the psychological facts of desire and purpose. What reason is there to exclude from this progression these highest of desires, these most exalted of aspirations? Indeed, their almost instinctive character seems to bring them even closer to the biological level than is the intellect itself. Thus to interpret the mysteries of the human spirit in the pedestrian terms of embryology may seem fantastic, but is it any more so than to believe that a thing of beauty can be broken down into a series of chemical reactions in its creator's brain? It is a lofty conception, I think, to regard the soaring spirit of man, which creates beauty, strives for knowledge, and aspires to an understanding of the mysteries of the universe, as rooted in the same vital processes which fashion his limbs and time the beating of his heart; to look upon the inspiration which welled up in Shakespeare's mind as he wrote *Hamlet* or in Beethoven's to find expression in the *Ninth Symphony*, or the imagination which pictured the "Last Supper" to Leonardo before he transferred it to the chapel wall, or the vision of St. Francis in the Portiuncula, as but loftier expressions of that same creative urgency that stirs in protoplasm everywhere. By means which still elude us but are the goal equally of the biologist, the poet, and the philosopher are born those yearnings which make man the noble animal he is. Living things are seekers and creators, and striving for goals is the essence of all life; but in man these goals have risen to heights before undreamed of, and he can set them ever higher at his will. Man's feet are planted in the dust, but he lifts his face to the stars.

The question still remains as to *why* these highest qualities of man came into being. The obvious answer is that they were useful to his survival and that individuals and races which possessed them had an advantage and were preserved by natural selection. Darwin's great generalization is still in high repute and may well explain the profound advances in man's reasoning powers which have been achieved since the days of the ape man, but selectionists have always had difficulty in explaining the origin of his spiritual qualities. To be stirred by beauty surely had no survival value in a primitive society, nor would a cave man who foggily began to ponder the mysteries of the world be likely to get his share of mastodon meat.

The ruthlessness of the struggle for existence has doubtless been over-emphasized, and many instances are known where there is cooperation rather than conflict between animals; but unselfishness and love of one's fellows seem so opposed to the very basis of natural selection that to derive them entirely by its means is to strain the theory. Too many times in the long history of man the light of civilization, kindled by desire for something higher than savagery, has been snuffed out by the barbarians. Even today, who will be bold enough to say that we are not still in peril from barbarism, from a savagery more refined but not less brutal than its paleolithic model? But always the upward tide, pushed back at one point, begins to pour in elsewhere; and slowly, despite discouragement and delay, it has continued steadily to rise. The selectionist will say that the very persistence of these higher ways of life proves their survival value. But it is not the selective elimination of barbarous and selfish individuals and societies which has lifted men from savagery. Civilization comes not from an improvement of the germ plasm but by experience and example, by the contagion of higher and more satisfying ways of life. The true cause, I believe, of man's upward climb is his persistent yearning for those values which to him seem higher and more satisfying and to which he instinctively aspires.

But whence do such strange longings arise? What is there that should make man crave these higher things? To the physiologist this is no mystery. These emotions, passions, and longings of the human spirit, whatever their evolutionary significance may be, must be anchored firmly in the chemistry of protoplasm, in the physiology of the nervous system itself. Is it not a fact that an extra supply of adrenalin, poured into the blood by the glands which secrete it, has a most profound effect on one's behavior? Many drugs are known which bring the user dreams ineffable. Alcohol can alter one's whole personality. Conscience dwells in the brain's frontal lobes, and, if these are severed, the patient need fear no more the chastening of this inner monitor. Genes control what one can taste and see and hear, and our judgment of what is beautiful must thus depend on our genetic constitution. Love between male and female is conditioned by the sex hormones. Even the tenderest of emotions, mother love itself, is dependent on a sufficient supply of prolactin in the blood. Here is our ancient problem once again. To tie the spirit to material things, to analyze a noble poem or symphony or picture into a series of chemical reactions and molecular changes, gives but a dusty answer indeed to him who would seek to understand the heights and depths of man's nature.

But if we are not satisfied with these somewhat pedestrian accounts of how man's spirit was born, what is there, short of mysticism, to which we can turn? One great fact in nature, it seems to me, does offer some light both for this deep question and for the general problem of goal-seeking. Man has climbed the age-long evolutionary stairway from its simplest beginnings. In this progress the organized living system in which successively his ancestral life was passed became vastly more complex, and the outreach of his mind and spirit grew ever wider. A strange paradox of nature is the contrast between this constant upward thrust in the evolution of life and the downward drift of all lifeless nature as pictured by the Second Law of Thermodynamics. This law, unshaken by all the upheavals of modern physics, is a prophecy of the fate of our material universe. It tells us that the higher forms of energy are being degraded to heat, which tends to spread itself evenly everywhere; that the randomness of things is continually increasing, and that complex physical systems tend always to be broken down to simpler ones; in short, that the universe is slowly "running down" to a dead level of uniformity. What wound it up is, of course, a major problem in cosmogony. But an equally significant question is the place of life in this vast process of material degradation; for life, in its evolution from simple and still unknown beginnings up to man, seems to move in just the opposite direction. This great drama shows a continual *increase* in complexity, a mounting tension, a steadily rising level of organization. "The fundamental thread that seems to run through the history of our world," says Needham, "is a *continuous rise in level of organization. . . .*" And again: "The law of evolution is a kind of converse of the second law of thermodynamics, equally irreversible but contrary in tendency."[8] There is here not an actual violation of the Second Law, for the energy by which life is maintained comes from the sun and is on its way down to lower levels; but the tendency and direction of change of the lifeless and of the living parts of nature are entirely different, a contrast which has been recognized by many. The course of evolution is marked by a continual rise in the level of the goals we have been discussing—developmental, physiological, and psychological—which life has set up and toward which it has moved. Just as the upward course of life introduces something new into the world of matter, so the emergence of these qualities of the human spirit has brought something new into biological evolution which finds no ready explanation in mechanisms by which the rest of the organic world has come to being. This is the culmination of that same goal-seeking process which is found in the simplest living cell. Pushing up against the weight of lifeless matter it has organized protoplasm into the specific bodily forms of plants and animals and expressed itself in desire, in

[8] Joseph Needham, *Time: The Refreshing River*, pp. 185, 230.

purpose, and in the triumphs of mind. Pushing up still further and seemingly beyond the governance of selective forces, it has flowered in the spirit of man and borne fruit in the lofty idealism which the noblest of our race have shown. Not only does life express itself in organized systems of exquisite complexity wherein goals at many levels are attained, but these systems are not static. Their goals are ever changing. Man rides the crest of this advancing wave. His nature and his destiny are ultimately those of life itself, and the longings of his spirit are part of the great upward surge of life from amoeba to man. Who knows how far it still may carry him?

This creative quality in life seems to be a unique attribute, setting it off from lifeless stuff; but what may be the origin of it we do not know. If the fate of life is not simply in the hands of outside forces but if the systems in which it exists move forward under their own power, so to speak, in directions governed by their own inner urgencies, this is a biological and philosophical fact of the first magnitude. If this inner directiveness, this autonomy, of life can be explored, we shall approach an understanding of the origin of those advancing goals in the human spirit, deriving them as part of life's unfolding course. Perhaps life has one great purpose, and the levels we have followed—developmental, physiological, psychological, and spiritual—may be successive stages in its ultimate fulfillment.

So much for the argument. The attempt in these

speculations to bring together toward solution the ancient problems of mind and body, of purpose, of value, of freedom, of the soul, and of the place of man's spirit in the universe by postulating for all of them a common basis in the fact of biological organization is surely as ambitious an undertaking as that to which a rash philosopher has ever laid his hand. For many readers my words must have carried scant conviction. I have certainly offended the materialists by bringing an element of mystery into what to them seems the straightforward story of the physico-chemical basis of life, uncomplicated by purpose, spirit, or other extraneous idea. Most biologists will not approve of mixing their science so thoroughly with philosophy, of complicating the discussion of organization and regulation by introducing overtones of psychology and metaphysics. Psychologists will surely regard the treatment of instinct, mind, and consciousness as far too simple and naïve. To men of faith, on the other hand, I shall seem to have surrendered at once to materialism by admitting that not only the psychical but even the highest spiritual qualities of man are all manifestations of the organizing capacity of protoplasm which shows itself also in the biological phenomena of bodily development and physiological regulation. To protagonists of both the ideological left and right I shall appear as a tepid compromiser, disliking to follow the logical necessities of science but afraid to take a firm stand on the side of the angels. I admit the validity of these strictures and can plead only the worthiness of my purpose. If what

I have attempted to do could be accomplished, if all the manifestations of life from the lowest to the highest could be gathered into a single bundle and shown to have an essential character in common, this indeed would satisfy our craving for simple interpretations and for unity in nature, and therefore it has seemed worth undertaking.

But more than this, such a treatment makes it possible to draw these great issues down to terms in which the biologist can talk with the philosopher about them and bring to their solution the great resources of his science. To regard them all as aspects of the problem of biological organization may seem a fantastic over-simplification, but at least it avoids beclouding the issue with a host of minor problems and gives us something very tangible to attack. The basic question is the origin and nature of this organizing, goal-seeking quality in life. Whatever we may think of the implications which I have tried to find in it, this at bottom is a perfectly definite biological problem with nothing metaphysical about it.

For this problem the materialist has his ancient answer ready: there is nothing here but the activity of a physicochemical mechanism, particularly complex but fundamentally no different from those with which we are already beginning to be familiar. Its self-regulatory character is essentially the same as that of an automatic machine, and to read into these activities any concepts useful in psychology or philosophy is but vague rationalizing, fit for acceptance only by those who believe in the stuff that dreams are made on. This is a defensible answer, and I cannot quarrel with it.

But it is not the only possible answer. For many of us the description it gives of man seems far too simple. If there arises in living stuff a goal, an image, a longing—call it what you will—which comes to expression in a noble deed, or a great poem, or a new insight into nature, does not this tell us something more profound than present scientific knowledge can do about that remarkable process which at its lowest level goes by the prosaic name of biological organization? This is not to advocate vitalism or any of the other subterfuges by which believers in the free spirit of man have tried from time to time to liberate him from the bonds of deterministic materialism. It is simply to suggest that there are in biology facts and principles as yet undiscovered which are concerned with this regulatory, goal-seeking, upward thrust of life. Perhaps a further study of bioelectrical fields will give a clue. Atomic physics, far as it now seems from biology, may help by the development of principles like that of Pauli, which already seems to have a bearing on the problem of organization; or that of Heisenberg on indeterminacy, which affects the whole question of freedom. Even telepathy and similar ideas should not be brushed aside as impossible, for they may well turn out to be significant for biology, as an eminent British zoologist, Professor A. C. Hardy,[9] has recently maintained. It is

[9] A. C. Hardy, "Zoology Outside the Laboratory," *Advancement of Science*, VI (1949), 213-23.

of the utmost importance to keep our minds open to suggestions from any quarter, however unpromising. Our program must be to push out vigorously across the frontier of the unknown everywhere and to explore that region where life, matter, and energy so mysteriously are joined. Surely science has a vast deal more to learn than now it knows. It will discover not only new facts but new concepts, new paths to understanding. "We fool ourselves," writes Hardy, "if we imagine that our present ideas about life and evolution are more than a tiny fraction of the truth yet to be discovered in the almost endless years ahead."

The plain fact is that in the present status of science biological organization remains still unexplained, and that many investigators are doubtful whether we are nearer to the ultimate answer than we were half a century ago. The problem may be fruitfully attacked in the laboratory by objective, experimental study of the regulation of form and function in plants and animals, bringing to this task all the resources which science has placed at our command. This is the highest goal of biology. When it is attained I believe we shall find that organization depends neither on the operation of only those physical laws which we now know nor on some superphysical or vitalistic agent about which nothing can be learned, but that a more perfect knowledge of nature and man will tell us how the physical and the spiritual are linked in that ascending, questing, creative system which is life. The answer may also be sought subjectively, I believe, in man's inmost experiences and intuitive perceptions, at biological levels far different from those of science. "A poem," says Vannevar Bush in his recent book, "can touch truths that go beyond those that are examinable by test tube or the indication of needles on instruments."[10] Life can be studied fruitfully in its highest as well as its lowest manifestations. The biochemist can tell us much about protoplasmic organization, but so can the artist. Life is the business of the poet as well as of the physiologist.

My argument is that, if the idealist will admit that *life* is his final problem and will halt his retreat to heights where the scientist is unable to follow him, he can successfully do battle at the level of biology itself and on its terms. Here he has the opportunity not only to defend himself but at last to counter-attack the position of his adversary. In this combat, let both opponents employ every scientific and dialectical force at their command to solve life's riddle, and agree to abide by the result. In that day when the verdict is finally rendered there will doubtless be surprises for both sides in store. The book of life then opened will prove to be the work not only of the biologists in laboratories but of those others who call themselves poets and artists, philosophers and men of faith, who yet all seek the same Promethean flame kindled in organized protoplasm of the humblest cell but rising thence to illuminate the world.

You will by now have guessed my own preference for this sort of aggressive idealism over the usual material-

[10] Vannevar Bush, *Modern Arms and Free Men*, p. 188.

istic position. Indeed, a thoroughgoing materialism of the sort that Haeckel advocated finds fewer supporters now. An understanding of modern physics and of relativity has convinced most thinkers that, as the old-fashioned, three-dimensional universe with its solid atoms and its Newtonian laws is out of date, so, too, is a philosophy which puts its trust in such a system. The problem is to bring the known and the unknown together in some satisfying way. The present hypothesis is an attempt to show how this may sometime be possible. Materialism and idealism doubtless will continue to be different ways of looking at the universe. Which will finally triumph, or what monistic philosophy may come from a merging of the two, no one yet can know; but the decision will be of the greatest moment to our race, for it will largely shape the character of that society which men will build.

You will share my disappointment that our excursion has not come to any certain answer for all of the great questions we so boldly faced at the beginning, questions "which, of old, men sought of seer and oracle and no reply was told." No sure solution, indeed, is to be found for them today. But if we can set these problems up against the background of life itself, if we can show that mind and body, spirit and matter, are held together in equal union as parts of that organized system which life is, then the idealist is encouraged to speak with much more confident voice. He can claim with assurance that mind is as real as body, for they are part of the same unity; that purpose and freedom are not illusions but are an essential part of the way in which events are brought to pass in protoplasmic systems; that the soul has a sound biological basis as the core of the integrated living organism; that our sense of values is not arbitrary but results from the directions and preferences shown by such systems; and that the course and history of life, so different from those of lifeless matter, give hope that it may have an inner directive quality of its own.

But if we have been less than successful in laying a foundation of solid biological certainty on which a philosophy for today can safely be erected, our speculation will perhaps encourage those who are not content to stay close to the safe shore of certainty but seek to launch out into the deep; who nourish what William James used to call "overbeliefs," rising still further into the unknown by faith but needing some assurance of firm fact beneath them. This we can help provide. The existence of an unsolved problem at the very core of biological science, and one which seems doubtful of solution by the concepts and principles now familiar to us, makes dogmatic materialism less assertive. Many of our overbeliefs cannot be proven true, but in a universe which still remains so far beyond our understanding they cannot longer be dismissed by the tough-minded as impossible and intellectually disreputable. Religious convictions and the philosophy of idealism have today a more respectful hearing than for many years.

If the goals set up in protoplasm, one thus may ask, the ends to which all living stuff aspires, have risen so high that in ourselves they now include the love of beauty and truth and goodness, may it not be that the organized system which man's spirit is, refined and elevated far above its simple origins, has grown to be the sensitive instrument through which he comes to recognize the presence of these same qualities in the universe outside him? A scientist[11] has well likened such qualities in us to "a hum given forth by the bronze bell of man as it catches a note from the eternal harmony and thrills respondingly from base to rim." Should we not look upon these qualities, to which our spirits can so readily be attuned, as lofty realities and worthy of our devoted loyalty?

If these responsive systems which we call our souls are found to be so stubbornly persistent in the flux of time and matter; if our personalities are each unique and seemingly so valuable in nature, does not this suggest that they may be of more significance, perhaps even of more permanence, than one would ever guess from a knowledge of the lifeless universe alone?

If these goals set up by the organizing power of life have been lifted ever higher through the ages, from simple protoplasmic patterns to the lofty aspirations of the human spirit, does not this bring a hope that life may be moving toward its own great goal and even that the universe itself is not the seat of aimless forces merely, of chance and randomness, but that it, too, has a pattern?

[11] Joseph Needham, *The Skeptical Biologist*, p. 40.

Is not faith an experience of the fundamental unity between our own highest goals and this great pattern of the universe?

If each of us is thus an organized and organizing center, a vortex pulling in matter and energy and knitting them into precise patterns; and if we are able, though in small degree, to create new patterns never known before, does not this suggest that we may actually be a part of the great creative power in nature and hold communion with it; and that, as James once said, we may come to recognize that this higher part of us is continuous with a *more* of the same quality operative in the universe outside and with which we can keep in working touch? Does not this, indeed, present as clear a picture as the scientist can draw of God Himself and our relation to Him?

The study of life—regulatory, purposeful, ascending—begins with protoplasm in the laboratory, but it can lead us out from thence to high adventure and to "thoughts beyond the reaches of our souls." In form of leaf and limb and in the beautiful coordination of their powers we see the first steps in that great progression which has long been marching upward from the first bit of living stuff toward some dim final goal, as yet but dreamed of, which the poet sings:

> "One God, one law, one element
> And one far-off divine event
> To which the whole creation moves."

SUGGESTED READINGS

The literature of this field is very extensive and is by no means covered fully by the following list of books. These are all in English or in good translation and may prove of interest to the reader who wishes to explore the subject further. Among them are represented the most important viewpoints in this controversial subject.

Bergson, Henri. *Creative Evolution.* Tr. by Arthur Mitchell. New York: Holt, 1911. 370 p.

Bertalanffy, Ludwig von. *Modern Theories of Development.* Tr. by J. H. Woodger. London: Oxford University Press, 1933. 204 p.

Compton, Arthur H. *The Human Meaning of Science.* Chapel Hill: University of North Carolina Press, 1940. 88 p.

Conklin, Edwin G. *The Direction of Human Evolution.* New York: Scribners, 1921. 247 p.

Driesch, Hans. *The Science and Philosophy of the Organism.* London: Black, 1908. 344 p.

Du Noüy, Lecomte. *The Road to Reason.* New York: Longmans, Green, 1949. 240 p.

Eddington, A. S. *The Nature of the Physical World.* New York: Macmillan, 1929. 353 p.

Haeckel, Ernst. *The Riddle of the Universe.* Tr. by Joseph McCabe. New York and London: Harper, 1900. 390 p.

Haldane, J. S. *Mechanism, Life and Personality.* New York: Dutton, 1914. 139 p.

Henderson, Lawrence J. *The Order of Nature.* Cambridge, Mass.: Harvard University Press, 1917. 230 p.

Hogben, Lancelot T. *The Nature of Living Matter.* London: Paul, Trench, Trubner, 1930. 316 p.

Holmes, S. J. *Organic Form and Related Biological Problems.* Berkeley: University of California Press, 1948. 163 p.

Huxley, J. S. *Man in the Modern World.* London: Chatto and Windus, 1947. 281 p.

Jeans, Sir James. *Physics and Philosophy.* New York: Macmillan, 1944. 217 p.

Jennings, H. S. *The Universe and Life.* New Haven: Yale University Press, 1933. 94 p.

Köhler, Wolfgang. *Gestalt Psychology.* New York: Liveright, 1920. 403 p.

Lillie, Ralph S. *General Biology and Philosophy of Organism.* Chicago: University of Chicago Press, 1945. 209 p.

Loeb, Jaques. *The Mechanistic Conception of Life.* Chicago: University of Chicago Press, 1912. 232 p.

McDougall, William. *The Riddle of Life.* London: Methuen, 1938. 273 p.

Millikan, Robert A. *Time, Matter, and Values.* Chapel Hill: University of North Carolina Press, 1932. 99 p.

Morgan, C. Lloyd. *Emergent Evolution.* New York: Henry Holt, 1923. 313 p.

Muller, Herbert J. *Science and Criticism.* New Haven: Yale University Press, 1943. 298 p.

Needham, Joseph S. *The Skeptical Biologist.* New York: Norton, 1930. 270 p.

———. *Order and Life.* New Haven: Yale University Press, 1936. 168 p.

Ritter, William E. *The Unity of the Organism.* Boston: Badger, 1919. 359 p.

Russell, Bertrand. *Religion and Science.* New York: Holt, 1935. 271 p.

Russell, E. S. *The Directiveness of Organic Activities.* Cambridge, England: Cambridge University Press, 1945. 192 p.

Schrödinger, Erwin. *What Is Life?* New York: Macmillan, 1947. 91 p.

Sherrington, Sir Charles S. *Man on His Nature.* Cambridge, England: Cambridge University Press, 1945.

Smuts, Jan C. *Holism and Evolution.* New York: Macmillan, 1926. 362 p.

Sullivan, J. W. N. *The Limitations of Science.* London: Chatto and Windus, 1933. 303 p.

Whitehead, Alfred N. *Science and the Modern World.* New York: Macmillan, 1926. 304 p.

Wiener, Norbert. *Cybernetics.* New York: Wiley, 1948. 194 p.

Woodger, J. S. *Biological Principles.* New York: Harcourt, Brace, 1929. 498 p.

INDEX